# TRANSNATIONAL MARRIAGES IN THE STEEL INDUSTRY

# TRANSNATIONAL MARRIAGES IN THE STEEL INDUSTRY

***Experience and Lessons for Global Business***

*Garth L. Mangum*

*Sae-Young Kim*

*Stephen B. Tallman*

Q

**QUORUM BOOKS**
Westport, Connecticut • London

**Library of Congress Cataloging-in-Publication Data**

Mangum, Garth L.
Transnational marriages in the steel industry : experiences and lessons for global business / Garth L. Mangum, Sae-Young Kim, Stephen B. Tallman.
p. cm.
Includes bibliographical references and index.
ISBN 1-56720-040-0 (alk. paper)
1. Steel industry and trade—United states—Mergers.
2. Consolidation and merger of corporations—United states.
3. International business enterprises—United States. 4. Joint ventures—United states. 5. Holding companies—United states.
6. Subsidiary corporations. 7. Competition, International.
I. Kim, Se-yong. II. Tallman, Stephen B. III. Title.
HD9515.M35 1996
338.8'3669142'0973—dc20 95-45964

British Library Cataloguing in Publication Data is available.



Library of Congress Catalog Card Number: 95-45964
ISBN: 1-56720-040-0

First published in 1996

Quorum Books, 88 Post Road West, Westport, CT 06881
An imprint of Greenwood Publishing Group, Inc.

Printed in the United States of America

The paper used in this book complies with the Permanent Paper Standard issued by the National Information Standards Organization (Z39.48-1984).

10 9 8 7 6 5 4 3 2 1

# Contents


Tables and Figures vii

**1 The Emergence of International Joint Ventures** **1**

The Magnitude of International Joint Venturing 2

Steel Industry Joint Venturing 2

Why Study International Joint Ventures? 3

Organization of the Book 7

Notes 7

**2 The International Joint Venture from a Theoretical Perspective** **9**

The Theory of Foreign Direct Investment 9

Defining the International Joint Venture 12

Joint Ventures and Other International Strategic Alliances 13

The Multinational Firm and the International Joint Venture 15

The Managerial Perspective on International Joint Ventures 19

Uses of Joint Ventures—Host-Country Perspectives 21

Expectations for International Steel Joint Ventures in the United States 23

Notes 23

**3 Steel Industry Challenges of the 1980s** 25
Steel on the Industrial Staircase 25
The Role of Steel in Economic Development 28
The World's Steelmakers 30
Back to U.S. Developments 39
Notes 60

**4 Steel Industry Joint Venture Responses** 63
Beginnings of the Steel Joint Venture Movement 63
Joint Venture Motives 66
Census of Steel Joint Ventures 67
The Continuing Thrust 71
Notes 74

**5 Steel Case Studies—Three Integrated Mills** 77
The National/NKK Joint Venture 77
Armco/Kawasaki 90
USS/Kobe 98
Notes 112

**6 Steel Case Studies—Three Finishing Mills** 115
California Steel Industries 115
A U.S./Korean Joint Venture 123
Inland and Nippon Steel 139
Notes 154

**7 Lessons from Steel Industry Joint Ventures** 157
The Steel Industry in 1995 157
Joint Venture Lessons 166
Back on the Industrial Staircase 175
Notes 177

**8 Theoretical Implications** 179
International Joint Venture Formation—Theory and Practice in the American Steel Industry 180
Managing IJVs in the American Steel Industry 189
Conclusion 195
Notes 196

Index 199

# Tables and Figures

## TABLES


| | | |
|---|---|---|
| 2.1 | Types of Cooperative Arrangements | 14 |
| 3.1 | Capacity versus Consumption for Major Countries | 38 |
| 3.2 | Adoption of New Technologies | 41 |
| 3.3 | U.S. Trade in Steel Mill Products | 45 |
| 3.4 | Hourly Wage Trends in the U.S. Steel Industry and All Manufacturing | 47 |
| 3.5 | Steel Industry Hourly Employment Costs | 48 |
| 3.6 | Labor Productivity for All Steel Industry Employees | 48 |
| 3.7 | Employment in the U.S. Steel Industry from 1973–1985 | 49 |
| 3.8 | Production Cost Consequences of Exchange Rates | 50 |
| 3.9 | Consolidated Return on Steel Industry Sales: Net Income/Sales | 53 |
| 3.10 | U.S. Comparative Return (Loss) as a Percentage of Equity for Integrated, Mini-Mill, and All Manufacturing from 1972–1983 | 57 |
| 3.11 | Output Levels, Costs, and Revenues for U.S. Integrated Facilities from 1976–1985 | 59 |
| 4.1 | Foreign Steelmakers' Investment in U.S. Joint Ventures | 68 |
| 5.1 | Production, Sales, Profits, and Losses, 1985–1993 | 86 |
| 7.1 | Production, Capacity, Utilization, and Income, 1984–1993 | 158 |

7.2 Percentage of U.S. Raw Steel Production by Furnace and Cast Type 159
7.3 Productivity Index 160
7.4 Employment and Labor Costs 161
7.5 Sources of Steel Supply 162
7.6 World Steel Production, 1984–1993 164
7.7 Hourly Compensation Costs for Production Workers 166
7.8 Income Data 167

**FIGURES**

3.1 The Industrial Staircase 26
3.2 Real Import Prices and Full Japanese Costs for Carbon Steel from 1973–1983 51
3.3 Real Domestic and Import Carbon Steel Prices from 1973–1983 56

# TRANSNATIONAL MARRIAGES IN THE STEEL INDUSTRY

CHAPTER 1

# The Emergence of International Joint Ventures

Globalization is generally perceived as a two-dimensional phenomenon. Firms in nations unable to sell their entire desired output within their own borders attempt to market the surplus abroad. There, consumers attracted by price or quality welcome these imports. At the same time, competing producers within those target nations attempt to erect protectionist barriers. Sometimes firms attempt to leap over borders to establish production outposts in target countries and abandon any concept of a home base, thereby becoming truly rootless multinationals.

Another international trade phenomenon has attracted less attention: that of transnational industrial marriages in which two competing business organizations maintain their respective home bases but join resources and combine strengths to exploit markets otherwise inadequately available to either. Such international joint ventures are the focus of this book, which illustrates the magnitude of the international joint venture movement, assesses its motivations, relies upon the basic steel industry for illustrations of its strengths and weaknesses, and draws lessons from that experience. Those lessons are the purpose of and justification for this book—strategic lessons for firms both within the steel industry and outside it, public policy lessons for those wondering whether to promote or discourage such international marriages, and academic lessons for those whose motives are only to understand and encase in theoretical simplifications the perplexing realities of a complex world.

## THE MAGNITUDE OF INTERNATIONAL JOINT VENTURING

By all accounts, the incidence of joint venture usage is climbing rapidly, both domestically and internationally. The speed of new technology emergence, the convergence of markets, and the high cost of new products argue for the value of cooperative arrangements. In the United States, Harrigan reports that the percentage of joint ventures and other cooperative strategies in relation to all organization-forming activity has increased dramatically across a variety of industries—from 1 or 2 percent before 1975 to between 10 and 35 percent of new ventures in more than half the categories she examines in 1985.[1] International alliances are comparably important. Contractor and Lorange, using Department of Commerce data, report that in 1977 there were some 10,000 foreign subsidiaries wholly owned by U.S. firms, some 14,000 to 15,000 equity participations, and 30,000 licensing arrangements.[2] The U.S. Department of Commerce reports that an average of almost 40 such foreign joint ventures and equity increases in the United States were counted annually between 1978 and 1983,[3] with a further 33 equity increases and 49 joint ventures in 1985.[4] Ellen Auster reports from Japanese government sources that Japanese firms formed 48 cooperative ventures and 33 equity joint ventures out of 136 total investments in the United States in 1984, and 76 cooperative ventures and 63 equity joint ventures among 223 total U.S. ventures in 1985.[5] In other cases, the incidence of joint ventures into China has increased from six in 1979 to 3,909 in 1988 and 3,659 in 1989 from all sources.[6] Other studies indicate a steady increase of cooperative ventures in Europe over the last decade, and comparable increases in a variety of industries.[7] Cooperation in general, and equity joint ventures in particular, are increasing in importance rapidly as firms deal with world markets and technology-intensive products.

## STEEL INDUSTRY JOINT VENTURING

Joint venturing in the world steel industry began in the years immediately following World War II, not in steel production but in the development of raw materials necessary to it. Iron ore, coking coal, limestone, and other more exotic necessities of the steelmaking process were not in short supply; but they were not always available in convenient locations and desired qualities. All-out war production had exhausted many of the most convenient and most attractive supplies and sent older producers and new entrants to the industry scurrying to find new sources, first in Canada in the 1950s; later in Australia, Africa, and Latin America; and later still in Asia.[8] Some of those international joint ventures were arranged between competing steel firms anxious to share the costs and the outputs, both of which tended to be beyond the capacities and needs of any one firm. Others were joint ventures between mining firms with both the expertise and the requirement of a guaranteed market and steel producers bringing to the marriage both the capital and the market.

Maturing of home markets during the 1960s accelerated the hunt for export markets during the 1970s in an atmosphere of rapidly rising energy and other costs. Primarily to integrate the raw-materials-supplying and steel-producing functions with consequent transportation-cost savings, new international joint ventures were undertaken by the Americans and Europeans in Australia, by Japanese and Brazilian firms in Brazil, and between Austrians and South Africans in the latter country. All foundered, either over government regulation or from the multiplying economic problems of the world steel industry documented in later chapters.

The political climate in the United States during those years was hostile toward any joint venture, foreign or domestic, that might be perceived as reducing competition. But "combine or perish" became the obvious alternatives for American steel firms during the 1980s, and antitrust policy responded to that reality. Superior quality had been demonstrated abroad which struggling U.S. producers could not obtain the capital to duplicate. Meanwhile, emerging world producers bursting the boundaries of their domestic markets and hungry to tap the world's largest and most lucrative market were willing and eager to offer capital, technology, and expertise.

The result was a one-industry joint venture boom that consummated at least seventeen international marriages in less than ten years.[9] Most of these are summarized, and six are examined closely in the chapters that follow.

## WHY STUDY INTERNATIONAL JOINT VENTURES?

International joint ventures may not have been the salvation of the steel industry in the United States, but that industry would be much smaller and substantially weaker today had those transnational marriages not occurred. Therefore, there are lessons for the industry to be reviewed. Those joint ventures helped change the look of steel in this country from an entrenched and protected but dying industry to a revitalized, technologically sophisticated, internationally competitive industry, lean but once again powerful. This impressive turnaround cannot be attributed entirely to the international cross-fertilization; but we shall see that foreign capital, foreign technology, and foreign commitment to steelmaking have been critical to saving plants, jobs, and even communities in the United States.

There is even more to be learned by industries of parallel and contrasting experience. Therefore, this book follows two paths in the investigation of international joint ventures in steel. First, it provides a historical narrative of the industry and the role of these international joint ventures in that industry's recent rejuvenation. By focusing on detailed case studies of six joint ventures, this book provides an in-depth look at why these ventures succeeded and failed; the practical value of joint ventures in this industry; the economics of foreign investment in one industry and country; and at the importance of human interaction in business relationships.

This aspect of the book is aimed at the manager with an interest in steel or international joint ventures. It shows what worked and did not work for these companies at a practical level in the hope of illuminating a complex process and providing ideas for other companies in similar situations. It is also intended for those who take a particular interest in the steel industry, whether serious scholars or casual readers who understand the importance of this industry to the industrial economy on both practical and symbolic levels. In this sense, this book provides a historical record (although one which is updated daily) because these ventures persist today; failing and succeeding, learning new lessons, and enriching their industry even as we try to evaluate what they did last month or last year. As such, this book is intended to be both entertaining and insightful. It provides six views of the real world of steel and suggests some conclusions about how these ventures were created and managed. It is also intended to provide the raw material with which readers can draw their own conclusions.

Another aspect of this book is that of a serious scholarly look at international joint ventures and their increasingly important position in today's economy. Alliance ventures of one sort or another have become an increasingly important form of international investment. Contractor and Lorange reported in 1988 that alliance forms of investment outnumbered wholly owned subsidiaries by more than four to one.[10] In concert with their increasing numbers, alliance strategies are becoming a favored means of investment in many industries: This is a considerable change from their previous status as a last resort when foreign governments blocked wholly owned investment. As outsourcing, re-engineering, and "redefining the firm" have become the hottest concepts in management technology, joint ventures and other alliances have exploded in both numbers and interest to managers and scholars of management. Their importance in global industry environments is at the top of the agenda.

A quick look at the changing environment of international business shows why this is an increasing trend. First, more and more industries are developing global markets. Global competition, world scale technology, and converging demands from customers in many lands force multinational firms to invest widely. Firms must challenge their competitors in their home markets and wherever they may find a competitive advantage. Whether competing for customers or for the latest technologies in product or process, or for the best value in input materials, companies face each other around the globe. New technologies are so expensive to develop, yet so fleeting in their superiority, that companies in electronics, pharmaceuticals, software, and other industries must have global distribution from the day of introduction of a new product. No longer can development costs be spread over years of technical superiority in a product life cycle that moves gradually from region to region and country to country. Now, product knowledge seems to spread almost instantaneously, and when customers everywhere want the same hamburger, computer, video game, or high-strength steel, the company that can first provide this worldwide standard reaps tremendous advantage from being first to match global tastes. Globalization is more than a cliched term, it is characteristic of the competitive situation in a growing number of industries, even the traditional "heavy industries" such as steel. In steel, new process and product technologies, internationalizing customers, and new competition put pressure on existing companies. This competitive trend which is critical, vastly expensive, and highly risky, provides one powerful incentive to pursue alliance strategies.

At the same time, new countries are moving into the world economy from the developing world and the old socialist world.

New political and economic alliances are developing, and old alliances are changing form. Countries previously dismissed as "less developed" or "centrally planned failures" are suddenly the fastest growing sources of both demand and supply. These countries provide hungry consumers and undervalued productive assets which cannot be ignored. Among the industrialized democracies, an era of alliances is also well under way: the European Union is proceeding with economic unification of a scope previously unknown among sovereign nations; the industrial nations of North America are committed to a free trade area; and efforts at economic cooperation in East Asia are beginning. These major changes in political and economic systems are paced by less globally significant, but equally wrenching changes in other parts of the world. International companies are forced to expand their geographical horizons to protect their access to markets and resources. Multinational businesses face a world where no borders are permanent, no political systems solid, and no government policies dependable. The uncertainty of the international business environment is perhaps as high as it has ever been without a major war, and the usual business response to uncertainty is to play safely and carefully—to hedge all bets.

How are multinational companies to deal with increasing uncertainty coupled with a burgeoning demand for global products, global technology, and global competitiveness? From a strategic perspective, greater uncertainty in the environment suggests careful thought, limited commitments, flexibility, diversification of risks, and protection of assets. Organizationally, such protective qualification of strategic direction is enhanced by the use of alliances in place of solo operations. Joint ventures reduce resource commitments in any one location, lowering diversifiable risks while permitting any single firm to cover more of the globe with the same assets. The internal difficulties and risks inherent to shared ownership and control of operations have caused alliance ventures to be treated as less than optimal organizational solutions for a long time. However, in times of turbulence and rapid growth, the risks from misjudging the direction of the environment, blowing an opportunity for market entry, or missing the introduction of a new technology appear to outweigh the risks of divided loyalties and leadership. When international firms must be able to operate effectively everywhere and anywhere while also remaining ready to make immediate adjustments in their worldwide systems, even the largest multinational companies need the resource leverage, the improved market access, the political connections, and the technological synergies that can be gained from alliances. Joint ventures make possible the bold but carefully hedged strategies needed to fit uncertain times.

## ORGANIZATION OF THE BOOK

Chapter 2 focuses on a discussion of conceptual and practical issues surrounding international joint ventures from both theoretical and managerial perspectives. This is followed in Chapter 3 by a description of the worldwide steel industry and its role in the economies of developing and industrial nations and a historical look at steel in the United States during the troubled quarter-century from the end of the 1950s until the mid-1980s. Chapter 4 provides an overview of the role of joint ventures and other alliances in the U.S. steel industry. This is followed by case studies of foreign joint ventures in the U.S. steel industry, three basic integrated mills in Chapter 5, and three finishing mills in Chapter 6. While the often assumed roles of national entities are reversed in these cases (most studies of foreign investment look at outward investment from America and other industrialized countries), we find that many of the strategic purposes behind these ventures are similar to those of more typical investments in joint ventures. Chapter 7 provides analysis of the events described in the case studies, drawing lessons from and for the steel industry in its strengthened condition nearing the end of the century. Chapter 8 returns to the scholarship of joint ventures, examining how these joint ventures in the steel industry can be placed with respect to the larger theories of management and economics, asking and answering again, what lessons can be learned?

## NOTES

1. Kathryn R. Harrigan, "Joint ventures and competitive strategy," *Strategic Management Journal* 9(1988): 141–158.

2. Farok J. Contractor and Peter Lorange, "Why should firms cooperate? The strategy and economic basis for cooperative ventures," in *Cooperative Strategies in International Business*, ed. F. J. Contractor and P. Lorange (Lexington, Mass.: Lexington Books, 1988), 3–28.

3. U.S. Department of Commerce, *Foreign Direct Investment in the United States: Completed Transactions, 1974–1983* (Washington, D.C.: U.S. Government Printing Office, 1985).

4. U.S. Department of Commerce, *Foreign Direct Investment in the United States, 1985 Transactions* (Washington, D.C.: U.S. Government Printing Office, 1986).

5. Ellen Auster, "The relationship of industry evolution to patterns of technological linkages, joint ventures and direct investment between U.S. and Japan," *Management Science* 38(6; 1992): 778–792.

6. Xiangming Chen and John W. Minton, "Foreign direct investment in the People's Republic of China." Paper presented at the Academy of Management conference, Miami Beach, Fla., August 11–14, 1991.

7. Michael Hergert and Deigan Morris, "Trends in international collaborative agreements," in *Cooperative Strategies in International Business*, ed. F. J. Contractor and P. Lorange (Lexington, Mass.: Lexington Books, 1988), 99–110.

8. William T. Hogan, *Global Steel in the 1990s: Growth or Decline* (Lexington, Mass.: Lexington Books, 1990), 205–207.

9. Martin Kenney and Richard Florida, *Beyond Mass Production: The Japanese System and Its Transfer to the U.S.* (New York: Oxford University Press, 1993), 157.

10. Contractor and Lorange, "Why should firms cooperate?" pp. 3–30.

CHAPTER 2

# The International Joint Venture from a Theoretical Perspective

This chapter explores the conceptual models of international joint ventures (IJVs), starting from both economic theory and from studies of the management of IJVs. By setting up the theory of joint venturing before addressing the specific cases of ventures in the U.S. steel industry, we hope to be able to put these cases in a broader organizational context and to learn from them about international joint ventures in general.

## THE THEORY OF FOREIGN DIRECT INVESTMENT

Before proceeding to joint venture theory, we should first take a quick look at foreign direct investment (FDI) in general. This allows us to approach our empirical study with three questions in mind. First, why would foreign firms enter the U.S. steel market? Second, why use foreign direct investment rather than exports to serve this market? Third, why use international joint ventures rather than building new plants or simply acquiring American firms? The answers to the first two questions will become apparent from the troubled recent history of the industry as discussed in Chapter 3. However, we can also place these questions in a theoretical model to gain some insight from wider experience.

The model of foreign direct investment which seems most appropriate to this situation is John Dunning's Eclectic Model of

FDI.[1] Dunning has written about his model extensively, and it has been discussed at length by other scholars. The summary provided here hardly addresses the nuances of the model, but will be sufficient to prepare for an exploration of the steel industry experience. Dunning begins by describing *ownership factors*, resources of a given multinational firm which provide a unique competitive advantage to that firm. He proposes that if the firm has no such advantages or competencies, it will not be able to compete in foreign markets given its inherent disadvantages of distance, shipping costs, tariffs, and unfamiliarity with the market. However, ownership advantage, in a perfect market and with the absence of other factors, only implies that the firm can successfully export to the foreign market.

In the steel industry, it is evident that foreign firms have had several advantages over U.S. firms. First, many of them have used more modern process technologies and were considerably more efficient than the American companies. Second, the quality of their product was often higher, particularly in the case of the Japanese steel companies. Third, due to process efficiencies, lower labor costs, and subsidies, foreign firms could sell profitably at lower prices. These advantages were reflected in the fact that despite transportation costs and delays and distance from customers, foreign imports were making significant inroads into the U.S. market in the 1970s and 1980s. The sustainability of such advantages, despite their seeming simplicity, is indicated by the rush of American steelmakers to leave the industry or seek government protection rather than compete directly. The institutional factors of domestic overcapacity, strong union pressures, and capital starvation kept the U.S. industry at a disadvantage even though it understood many of the causes for its problems.

The foreigners were able to make better, cheaper steel and export it to the United States at a profit. Why then would some of them decide to move production to the United States? Dunning describes *location factors* which determine *where* production will take place in an international business. These factors provide advantage to the firm by locating its operations in that place whether home or host country (or, possibly, even a third country) and where the economics of the location provide the greatest efficiencies. In the early stages of internationalization of the steel industry, raw materials were traded from places that had them to places where they were needed by the market. Basic trade economics and comparative advantage explain this. Finished steel first began to be traded because the input factors of production, primarily labor, were much less expensive in some locations, particularly Japan and Korea, than in the United States, the primary importing nation. In other cases, especially that of western European producers, government subsidies lowered production costs and encouraged exports to maintain employment levels in the industry. Thus, government distortions of the market encouraged some trade in steel. Further threats of government intervention, this time on the part of the U.S. government, were a primary encouragement for foreign producers to move operations to the United States.

Steel industry pressure resulted in a number of anti-dumping actions against foreign producers while threats of punitive and protective tariffs loomed large as the U.S. industry went through a period of shrinking production, layoffs, and plant closures. At the same time, labor cost advantages began to disappear for the Japanese and primary customers, also responding to U.S. protectionist threats (primarily in the auto industry), and moved to American-based production. Increasing home costs, a customer base which had shifted to the United States, and various real, threatened, and anticipated trade restrictions made production in the United States begin to appear to make economic sense. Recognition that their product and process technologies were perhaps superior to U.S. practice moved the competitive advantage of foreign producers away from the home country, location-based, low-factor costs to company-based technological competencies which could be transplanted.

The final step toward direct investment in Dunning's model is the recognition of *internalization factors*, organizational capabilities which make ownership and direct control of overseas operations preferable to market controls. Of primary interest are organizationally bound, complex skills which are difficult to fully describe in documents or involve complex social interaction in teams. Such tacit resources encourage the firm to become multinational by extending its internal organization abroad. Through foreign direct investment, the company is able to accomplish two main goals. First, it can protect its proprietary knowledge from compromise to a licensee or export sales agency. Many companies have discovered that training partners in applying their unique skills, without ownership, has led to creating competitors while providing inadequate royalty compensation. Second, the firm can ensure that its unique capabilities and resources are applied properly. Thus, a valuable brand name is not ruined by being attached to low-quality goods or poor service if a unique product technology is engineered correctly and a complex production process is organized and managed correctly. The multinational company can both protect its sources of competitive advantage and manage them for maximum economic benefit. The Eclectic Model provides a reasonable explanation, from an economic perspective, for foreign direct investment, but provides no direct explanation for joint ventures as a specific form of FDI. That is the subject of the following sections.

## DEFINING THE INTERNATIONAL JOINT VENTURE

As we intend to focus on international shared equity joint ventures, we should first define in detail just what we mean by this categorization. John Kwoka defined joint ventures as the integration of operations between two or more separate firms, in which the following conditions are present:

1. The enterprise is under the joint control of the parent firms, which are not under related control.
2. Each parent has invested a substantial amount in the joint venture company.
3. The joint venture exists as a business entity separate from its parents.[2]

There are two general forms of organization which meet this definition. One is the acquisition of *part* of one firm by another. In this case, one partner or owner is part of a distinctly separate firm, while the remaining ownership is shared among whatever private or public equity holding existed previous to the partial acquisition. The other form is the establishment by two or more independent companies of a third business entity in which the parents each own significant equity shares. Here, the parents share ownership, management responsibilities, and profits of the joint venture company. The parents may hold equal shares of the joint venture, or they may participate as majority and minority interests.

IJVs are further defined as shared equity organizations in which the joint venture and at least one of the parent firms are based in different countries.[3] Most IJVs conform to the model of a multinational parent and a local parent or group of local equity holders for which a primary motivation is access to the local economy. However, other IJV ownership structures may include two parents from one home country organizing a joint venture in a foreign country where their combined resources appear to offer unique advantages either for market access, raw material access, or offshore production. Also, parents from two or more foreign countries may set up a strategic alliance in a third country in order to apply their pooled assets to that market.

Many academic and managerially oriented books and articles have been written about joint ventures and other strategic business alliances. Equity joint ventures are but one variety of strategic alliance, with unique advantages and disadvantages over other forms. In order to study IJVs in their full organizational context, we need a common understanding of their role in the world of alliances. The next section provides a summary of the alliance forms which places the equity joint venture on the continuum of alliance organizational forms.

## JOINT VENTURES AND OTHER INTERNATIONAL STRATEGIC ALLIANCES

A variety of terms have been used to describe cooperative relationships among business firms. *Strategic alliance* is a broad term encompassing most of the cooperative forms. It also implies the significance of cooperative forms to the strategic success of the companies in question. No longer are long-term licensing agreements or equity joint ventures considered as second best responses to government requirements or exotic secondary markets. Rather, cooperative arrangements are used to change the terms of competition across major industries. The subtle differences among the various types of alliances make strategic nuance possible. An extensive list of possible alliance types is provided by Farok Contractor and Peter Lorange[4] and is shown in Table 2.1.

Table 2.1 indicates the variety of alliances, listed in order from the least permanent and lowest commitment extended license agreements to equity joint ventures. Technical training or start-up assistance agreements require involvement of the two organizations over the period of time which covers the training period and can be terminated by prearrangement. The degree of involvement is defined by that which is needed to support a small training team in a foreign location for what is usually a short time. Franchising predicates a long and indeterminate lifetime for the alliance and requires only limited commitment. Most franchise agreements are detailed, requiring only periodic monitoring. Enforcement is primarily in the form of agreed-upon termination for non-performance.

Research partnerships and other high level agreements anticipate long-term alliances with significant degrees of interaction between the partners in a manner difficult to fully specify,

**Table 2.1**
**Types of Cooperative Arrangements**

| Type of Cooperative Agreement | Extent of Integration |
|---|---|
| Technical training and start-up assistance agreements | Low |
| Production, assembly, buy-back agreements | ↓ |
| Patent licensing | ↓ |
| Franchising | ↓ |
| Know-how licensing | ↓ |
| Management and marketing service agreements | ↓ |
| Partnerships in<br>Exploration<br>Research<br>Development or coproduction | Moderate<br>↓ |
| Equity joint venture | High |

thus engendering significant organizational commitments. Such alliances subject the partner firms to mutually high potential risks from opportunism or cheating by the other firm. The most significant degree of partner interaction in an alliance is shared equity ownership in a joint venture.

Stock ownership permits the partners freedom to access information, monitor performance, and control operations in ways which would be subject to high risks and intense negotiations in a contractual mode. Partners can protect their share of the residual income and have input to decision making without constant renegotiation. Equity ventures, by setting up new organizations, can increase the loyalty of managers, who are removed from the parents, and workers, who are usually hired directly by the joint venture, and can encourage transfer of organizationally embedded, implicit knowledge and group skills.

At the same time, shared equity joint ventures do pose risks to the alliance partners. The partner firms may gain new knowledge from their ally, but they also risk opening themselves to the same "freedom-of-information" access by the ally. The flexibility and managerial discretion in equity joint ventures make possible the sharing of complex organizational knowledge and skills as well as the opportunity for managers to run the joint venture as an independent firm. This freedom puts the know-how of the parents on the table to some extent, but also provides much greater potential for sustained relationships and high levels of profitability. The complexity of the interactions in shared equity joint ventures makes them of particular interest to alliance studies. This series of case studies focuses on equity joint ventures in the steel industry, both from the importance of the industry and the revelations about the globalizing world economy to be found in these ventures. The next section discusses the theoretical and managerial issues behind the use of IJVs from the perspective of the multinational firm. This is followed by a discussion of the impact of joint ventures involving foreign firms on the host nation economy.

## THE MULTINATIONAL FIRM AND THE INTERNATIONAL JOINT VENTURE

Joint ventures are claimed to be strategic responses to uncertainty and growth in the environment. How does this work? Theoretical answers are provided by the major economic models of the multinational firm.[5] The *Strategic Behavior Model* of the multinational firm, based on the concepts of industrial organization economics, suggests that strategic alliances permit large oligopolistic firms to extend their influence widely, even in the face of government resistance and competition. Such firms identify and occupy market positions which can be defended against further entry, thus permitting them to gain and share monopoly profits. Internationally, this type of strategy suggests FDI strategies which allow the multinational firm to lock up foreign markets. When wholly owned subsidiaries are not permitted, joint ventures provide a long-term market position through association with a host-country firm. Potentially threatening local firms can be absorbed or otherwise controlled by provision of technology, skills, and capital so that they aid the multinational firm rather than attacking it.

An alternative interpretation of the reasons for joint venture formation is provided by the *Internalization Theory*, an international version of transaction cost economics.[6] This model suggests that an alliance, and a joint venture in particular, acts as a performance bond when two firms wish to combine resources yet do not trust each other to act fairly in a pure-market transaction and do not desire a full merger of assets. Hennart proposes that when two firms each possess unique resources or capabilities which would have higher values if combined but for which market prices cannot be determined, a joint venture may be the answer.[7] Joint ownership allows the firms to combine even difficult-to-transmit capabilities without having to deal with the large numbers of irrelevant assets which would be included in an acquisition. The parents can then monitor the use or misuse of the assets and skills which they have each provided. Long-term equity alliances guarantee that both sides can monitor the transaction and that rewards to both partners are mutually related to the success of the venture. If either partner cheats, the joint venture will be harmed, costing both equity holders a part of their earnings. Lack of confidence in a fair deal or a reliable alliance partner is exacerbated by the different outlooks of managers in firms from different countries, making joint ventures particularly likely in international dealings.

A different but compatible perspective on IJVs is provided by the resource-based economics view of the firm, which focuses on bundling strategic and generic resources to gain *economic rents*, or profits.[8] This view of businesses suggests that joint ventures allow firms to combine valuable resources which are related to tacit know-how and are tied broadly to the organization and its routines. Such competencies are impossible to transmit through licenses or patents because they involve team activities and learning by doing.[9] Therefore, sharing such complex skills requires transplanting part of the organization and its people so that the alliance can learn how to perform the requisite tasks from both parents. While combining such implicit skills requires the close and extended relationships of a joint venture, the parents risk the loss of their particular core competencies to the joint venture and to their partner, particularly if one is more aggressive about learning than the other.[10] Chi shows that forming these bundles of strategic and supporting, or complementary, assets involves non-recoverable transactional investments which make market, alliance, and wholly owned packages non-equivalent.[11]

Porter says that different environments in different home countries lead to firms in these countries evolving distinctively different capabilities.[12] The organizational skills most likely to develop in this way are tied so closely to the firm itself that they cannot be transmitted to a firm from a different nation. Thus, the unique environments of firms from different home countries both encourage bundling complementary skills and make international equity alliances the best way to meld these embedded competencies.

Resource-based and transaction cost arguments for joint ventures support one another. The resource-based model explains why the profits available to the joint venture may exceed market or ownership relationships—the bundling of two or more sets of unique organizational resources to produce larger joint profits. Transaction costs explain the value of using an equity alliance—protecting the parent company's interest in the skills and assets which it shares with the joint venture while making the knowledge transaction as efficient as possible. The projections of both theories are enhanced in the international arena, where firms have very different capabilities and less understanding of each other.

Other economic models provide insight on joint venture formation and management. Evolutionary economics suggest that the joint venture will provide an opportunity to try new combinations of capabilities, some of which will provide new economic value to the joint venture and to one or both of the parents. The joint venture thus becomes a learning opportunity for the parents, both through observing the partner's capabilities in action and by trying new ways of applying one's own competencies. Finally, the use of game theory in economic modeling shows that while competitive relations can be preferred in a single iteration game, cooperation provides the best outcome for both players in a multiple turn game, particularly with non-zero sum payoffs.[13] If alliances are treated as true sources of advantage, such that internal opportunism is eliminated as non-productive, stable cooperative relationships of business organizations appear to have great theoretical potential as strategic solutions to the uncertainties and demands of a dynamic but confusing environment which is typical of international conditions.

Organizational theory also addresses joint ventures as a form of business organization. Shenkar and Tallman suggest that a number of organizational considerations can lead to specific forms of alliance.[14] Pfeffer and Nowak show that firms act to reduce the potential ill-effects of excessive dependency on other firms through joint ventures.[15] Internal compromise in approaching overseas markets and unfamiliar countries, partial interdependencies among the interests of two firms, and corporate culture and institutional routines influence the selection of strategic alliance type for a multinational firm. The similarities and differences of national cultures between home and host countries will influence the choice of organizational form in specific ventures, and the requirements, needs, and desires of other concerned stakeholders (government, labor unions, investors, etc.) in a particular host location will also affect the choice of alliance form. Traditional international business studies of IJV activity emphasize the importance of cultural distance as an incentive for the hedged approach to the market represented by the IJV.

Institutional models suggest that uncertainty leads managers to be conservative in committing resources to new directions, encouraging joint venture use in new or unfamiliar markets, particularly when joint ventures are common in an industry.[16] Kogut and Singh show that joint ventures are preferred over acquisitions when the cultural distance between home and host country is greater, thus creating uncertainties.[17] Kobrin shows that among less-developed regions, American multinationals use more shared control because host regions are farther from the United States, both geographically and culturally.[18] Various studies suggest that partners with complementary task-related assets and related firm cultures form more lasting alliances.[19] On the other hand, Arvind Parkhe, proposes that while complementary economic resources provide incentive to form joint ventures, dissimilar company cultures are the most common reason for IJVs to dissolve.[20] The different national cultures influencing the firms of different national origin in an IJV arrangement make such an outcome quite likely.

In summary, economic theory provides a variety of conditions which might encourage joint ventures. However, economics can only say that failure indicates a change in the conditions that favored joint venturing, resulting in a shift to another form. Organizational models, on the other hand, provide less convincing arguments for why IJVs might be a preferred form, but are much more convincing in explaining why they are a hedge against uncertainty (a second-best option) and why they so often fail to last. When cultural clash is included in the picture, the simple economics of complementary resources seem rather naive. Fear of differences and incompatibility of interests provide an equally convincing, somewhat more intuitive, and distinctly less appealing view of the use of joint ventures in international markets.

## THE MANAGERIAL PERSPECTIVE ON INTERNATIONAL JOINT VENTURES

Theoretical considerations suggest that uncertain environments may generate great numbers of joint ventures, and that such alliances may provide a superior outcome for the participants if they are accepted as a preferred strategy. How do such considerations appear in the real world of international business, where the political and economic conditions described above are a fact of daily life for managers?

Joint ventures, when considered as alternatives to acquisition and whole ownership, offer access to critical assets or markets with a minimum of resource commitment and increased flexibility with lower opportunity costs. Hergert and Morris, in a study of alliances among firms from industrialized countries, found that some 74 percent of alliances involved European partners, and that the rate of alliances in Europe was increasing.[21] An article in *The Economist* from June 1990 also showed an increasing rate of equity joint venture formation in western Europe. Similar findings by other studies suggested strongly that firms were responding to the prospect of the 1992 economic union of the European Community nations through strategic alliances. Hergert and Morris also found that most alliances were formed in technology-intensive industries, where high development costs and short product life cycles are forcing firms to enter many national markets with inadequate preparation, and that the majority of the alliances joined two rivals. The overall picture is one of strategic response to a vital new market, the exact conditions of which were difficult to define, so that alliances permit presence without excessive resource commitments to any particular course of action.

Recent investments in eastern Europe are outlined in *Bloc* magazine, January–February 1991. This article suggests that the largest investments in this area are for the purpose of providing consumer goods in newly developing markets. PepsiCo and General Motors, for instance, are investing in joint ventures to provide fast foods and auto parts, respectively, to Russian consumers. In other cases, such as Ford in Hungary or Anglo–Suisse–Phibro in Russia, investments are intended to stimulate exports, earning hard currency for both the multinational and the local partners. Joint ventures make operation in a new, very different environment less daunting for Western firms. The motivations of potential customers and potential workers newly released from decades of centrally planned socialism can seem quite mysterious to companies steeped in open-market traditions. A local partner can provide the guidance needed to make the risks of such emerging-market investments seem reasonable, even as the governments of these states have eased regulatory requirements for local partnering.

Shenkar describes the impact of environmental and organizational issues on alliances in China, where cooperative forms far exceed wholly owned ventures.[22] He shows that environmental differences between the multinational partner's home country and China are related to the use of non-equity alliances. His main finding, however, is that conflicting goals on the part of the foreign partner and the Chinese government reduce commitment and increase the chances for failure. Such conflicts include, for instance, the Chinese desire for management training versus the multinational firm's desire for minimizing labor costs, as well as the Chinese desire for technology competing with the multinational's desire to protect its patents and know-how. While whole ownership is possible and joint ventures have these risks, lack of knowledge about the Chinese culture, people, market, economy, and political system combined with an inability to access needed inputs without local connections make joint ventures ubiquitous in China.

These few examples suggest that the uncertainty connected with the unification of western Europe and that associated with the freeing of markets in eastern Europe and China are generating increasing alliance activity. The emphasis of most writings is on equity joint venture formation, but other forms of alliances are increasing as well. Cooperation among firms from industrial countries in high-technology, or capital-intensive industries may be most often characterized as non-equity. Limited contractual alliances permit sharing necessary technologies while protecting other important secrets.[23] In joint ventures among multinationals, while careful attention to management, monitoring of performance and operations, and specific goals are needed, the similarities in size and goals of the partners make alliance a natural, if guarded, relationship.

In many situations, however, equity ventures with their broader roles and less restricted inputs may be needed. While providing advantages of market access, cost reduction, distribution channel access, and competitive positioning, alliances in newly opening nations are affected by foreign exchange problems, limited local management skills, and underdeveloped infrastructures.[24] Such problems are impossible to address in contractual language, requiring the flexibility and commitment of an ownership interest and active management control. Therefore, we would expect greater success with the admittedly higher risk alliance types that involve more commitment from the partners. If partners are chosen with care and the alliance is developed with skill and patience, the potential risks of opportunistic behavior appear to be small in practice.[25]

In other cases, the information being shared is simply too complex and tied to the parent organization to be transmitted in a document. People and know-how must be shared. This is the situation with much of the steelmaking technology which is the focus of the cases in this study. While the technologies can be described formally, actually applying them in the real world of the steel plant involves levels of experience and understanding that cannot be communicated except through direct contact. Joint ventures or partial acquisition, with their shared-management control, exchange of employees, flexibility, and general depth of association are designed for just such situations.

## USES OF JOINT VENTURES—HOST-COUNTRY PERSPECTIVES

While joint ventures may involve two or more multinational firms (perhaps in a market which is native to none of them), the majority of alliances still involves a local host-country firm. This local firm may itself be a large multinational company, or it may not be; but in the case at hand, it is valued more for its location than for its other assets. Foreign partners seek strategic alliances for purposes of eased market entry, reduced financial risk, and vertical integration. However, the host country, or local firm is not just a target for multinationals to shoot at, but is often an active seeker of alliances itself. In such cases, what advantages are there for the local partner to make such an alliance desirable?

The most commonly identified advantage which is sought by local firms in forming strategic alliances with foreign multinationals is newer, better technology or other organization-specific knowledge.[26] This is true in China,[27] in Poland,[28] and in most other less developed regions. It is also the primary objective of the U.S. steel companies in the cases studied here. Technology development and exploitation is at the heart of resource-based and transaction cost models of joint ventures. A second major objective of local parents is access to capital and hard currency. Capital is important to host-country firms in less developed nations, but may also be an issue for ventures in industrialized nations. For instance, Japanese research partnerships with small U.S.

firms in high-technology industries such as microelectronics or biotechnology often involve the exchange of Japanese capitalization for American technology. Although the United States has large supplies of capital, investment in steel during the period studied here was less than popular. Steel companies often sought diversified investment, and financial markets had little interest in a "rust-belt" industry with excess capacity, old technology, and heavy foreign competition.

Foreign parents may also provide access to international markets in much the same way that the local parents provide entry to the host country market. Chinese ventures, in particular, often rely on the foreign partner to buy back a certain percentage of production for sales abroad in exchange for limited access to the Chinese market. On the other hand, the foreign parent may provide access to cheaper or better inputs for the productive process from its greater economies of scale in upstream production, access to cheaper raw materials or labor, or from its superior process technology. Access to high value raw materials for the U.S. joint ventures was a critical input from the foreign parents in the steel industry cases illustrated here.

The local partner, much like the multinational partner, seeks access to some resources or skills which are difficult or expensive to transmit in a one-time market transaction, but under conditions where whole ownership is not desirable. Typically, the local firm cannot buy out the multinational, and it fears loss of identity if it is acquired. Indeed, since the key resources of the local firm are so often tied to location requiring no specific investment which must be repaid, the risks of alliance are low for the local firm. The multinational partner may fear giving away its proprietary technology, but a firm from the host country cannot lose its national identity. Should the alliance eventually end, the host firm will have had the advantage of a strong partner for a time, and may be able to use this relationship to build a strong future position. For the same reason, we would expect the host-country partner to favor equity joint ventures, in hopes of gaining knowledge beyond the strict limits of the formal alliance while risking little. Strictly drawn contractual alliances offer few opportunities to gain access to additional technology or know-how, but ownership rights may open the joint venture to unexpected information. In addition, equity participation provides capital directly where contractual alliances typically do not. When such fungible resources are needed by the local partner, giving up part ownership may seem quite reasonable. If the foreign partner later wishes to up its stake, a fair or overstated market price provides the local investors with compensation that a terminated contract does not.

## EXPECTATIONS FOR INTERNATIONAL STEEL JOINT VENTURES IN THE UNITED STATES

This chapter has presented both theoretical and pragmatic arguments for the value, and the risks, of IJVs. The following chapter will set the contextual stage for the IJVs which are presented in detail in subsequent chapters. If we hope to understand the use of IJVs in steel better, and to use these examples to understand a bit more about the bigger picture of IJV theory, we need apply the general models described in this chapter to the specific contexts of these cases.

## NOTES

1. John Dunning, "The eclectic paradigm of international production: A restatement and some possible extensions," *Journal of International Business Studies* 19(1; 1988): 1–32.

2. John E. Kwoka, Jr., "International joint venture: General Motors and Toyota," in *The Antitrust Revolution*, ed. John E. Kwoka, Jr., and Lawrence J. White (New York: HarperCollins, 1989), 47–49; Joseph Brodley, "Joint ventures and antitrust policy," *Harvard Law Review* 95(May 1982): 1521–1590, quoted in *The Antitrust Imperative*, ed. John E. Kwoka, Jr., and Lawrence J. White (New York: HarperCollins, 1989).

3. J. Michael Geringer, "Strategic determinants of partner selection criteria in international joint ventures," *Journal of International Business Studies* 22(1; 1991): 41–62.

4. Farok J. Contractor and Peter Lorange, "Why should firms cooperate? The strategy and economic basis for cooperative ventures," in *Cooperative Strategies in International Business*, ed. F. J. Contractor and P. Lorange (Lexington, Mass.: Lexington Books, 1988), 3–28.

5. Bruce Kogut, "Joint ventures: Theoretical and empirical perspectives," *Strategic Management Journal* 9(1988): 319–332.

6. Peter J. Buckley and Mark Casson, "A theory of cooperation in international business," in *Cooperative Strategies in International Business*, ed. F. J. Contractor and P. Lorange (Lexington, Mass.: Lexington Books, 1988), 31–54.

7. Jean-Francois Hennart, "The transaction costs theory of joint ventures: An empirical study of Japanese subsidiaries in the United States," *Management Science* 37(4; 1991): 483–497.

8. Edith Penrose, *The Theory of the Growth of the Firm* (New York: Wiley, 1959).

9. C. K. Prahalad and Gary Hamel, "The core competence of the corporation," *Harvard Business Review* 68(May–June 1990): 79–91.

10. Gary Hamel, "Competition for competence and inter-partner learning within international strategic alliances," *Strategic Management Journal* 12(Special issue; 1991): 83–103.

11. Tailan Chi, "Trading in strategic resources: Necessary conditions, transaction cost problems, and choice of exchange structure," *Strategic Management Journal* 15(4; 1994): 271–290.

12. Michael E. Porter, *The Competitive Advantage of Nations* (New York: Free Press, 1990).

13. Arvind Parkhe, "Strategic alliance structuring: A game theoretic and transaction cost examination of interfirm cooperation," *Academy of Management Journal* 36(4; 1993): 794–829.

14. Stephen B. Tallman and Oded Shenkar, "A managerial decision model of international cooperative venture formation," *Journal of International Business Studies* 25(1; 1994): 91–114.

15. Jeffrey Pfeffer and Phillip Nowak, "Joint ventures and interorganizational interdependence," *Administrative Science Quarterly* 21(1976): 398–418.

16. Paul DiMaggio and Walter W. Powell, "The iron cage revisited," *American Sociological Review* 48(2; 1983): 147–160.

17. Bruce Kogut and Harbir Singh, "Entering the United States by acquisition or joint venture." Reginald H. Jones working paper series presented at the Wharton School, University of Pennsylvania, 1986.

18. Stephen J. Kobrin, "Trends in ownership of U.S. manufacturing subsidiaries in developing countries: An interindustry analysis," in *Cooperative Strategies in International Business*, ed. F. J. Contractor and P. Lorange (Lexington, Mass.: Lexington Books, 1988), 129–142.

19. Geringer, "Strategic determinants."

20. Arvind Parkhe, "Interfirm diversity, organizational learning, and longevity in global strategic alliances," *Journal of International Business Studies* 22(4; 1991): 579–602.

21. Michael Hergert and Deigan Morris, "Trends in international collaborative agreements," in *Cooperative Strategies in International Business*, ed. F. J. Contractor and P. Lorange (Lexington, Mass.: Lexington Books, 1988), 99–110.

22. Oded Shenkar, "International joint ventures' problems in China," *Long Range Planning* 23(3; 1990): 82–90.

23. John Pattison, "Global joint ventures," *Overseas Business* (Winter 1990): 24–29.

24. Oded Shenkar and Krystzof Obloj, "International joint ventures in Poland." Working paper presented at Tel-Aviv University, 1991.

25. Christopher A. Bartlett and Sumantra Ghoshal, *Transnational Management* (Homewood, Ill.: Irwin, 1992).

26. Rebecca I. Porterfield and Michael J. Stahl, "Joint venture decisions," in *Strategic Executive Decisions*, ed. M. Stahl (Westport, Conn.: Quorum Books, 1989).

27. Shenkar, "International joint ventures' problems."

28. Shenkar and Obloj, "International joint ventures."

CHAPTER 3

# Steel Industry Challenges of the 1980s

The experience of the world steel industry can be particularly instructive in understanding the reasons for and the results of IJVs. Not only has steel been the locale of a high proportion of such ventures, but the industry's key role in the economic development process has made the lessons drawn extraordinarily meaningful.

## STEEL ON THE INDUSTRIAL STAIRCASE

Looking back over the past two centuries, it is useful to perceive the path of economic development as steps up an industrial staircase (Figure 3.1).[1]

Historically, the few have generally been able to grasp the good life from the many, whatever the state of prosperity. But creature comforts become available to the majority only if productivity climbs, usually by improved skills and technology—the driving forces of industrialization. It is pursuit of that rising productivity which generally drives nations and peoples up that industrial staircase. Industrialization generally begins when productivity in the primary and extractive industries of agriculture, fishing, and forestry rises above the subsistence level, freeing labor from the daily grind of producing just enough food to

**Figure 3.1**
**The Industrial Staircase**

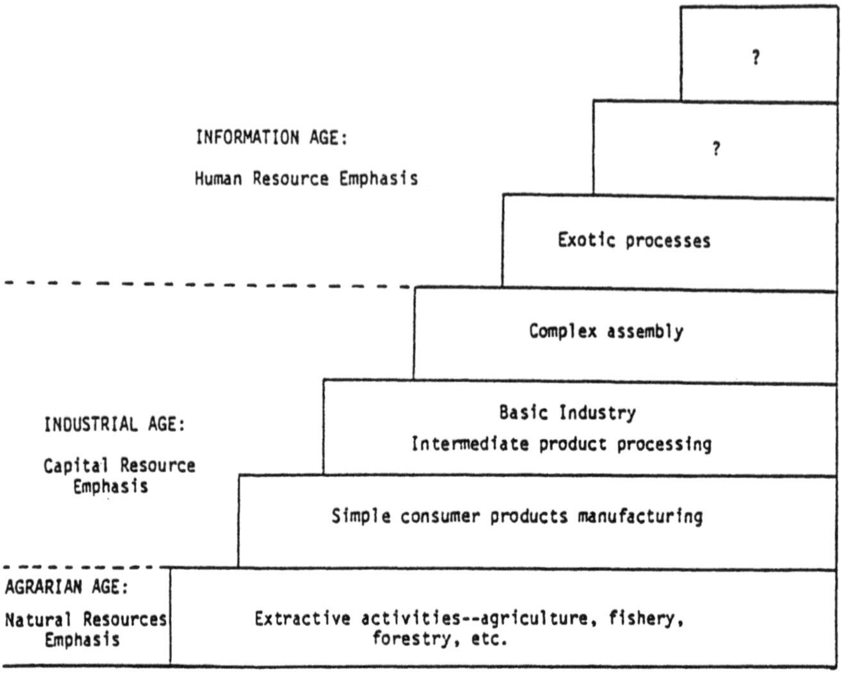


stay alive and allowing simple manufacturing activities to enrich life and further increase productivity.

Mining as an extractive industry, with the exception of precious metals, is more likely to be the handmaiden of manufacturing. The latter begins with simple products and simple production processes which were previously homemade or imported, labor intensive, and doable by unskilled labor with minimal equipment, and serving accessible domestic markets. Traditionally, that first step up on the industrial staircase was the manufacturing of textiles, clothing, and shoes. In modern times, basic electronic assembly, farmed out from already developed countries, tapping the cheap labor of nations not yet prosperous enough to provide a market for what they are producing, has been added to that first step. Thailand, Malaysia, and Indonesia might serve as current examples.

This first step on the industrial staircase offers profits and wages above those of farming and fishing, but it is still limited by its labor intensiveness. The use of relatively simple machinery is the key to both profits and the ability to pay higher wages. The second step, therefore, is into more capital-intensive industries that process available natural resources into intermediate products and require movement up the skill ladder to primarily semiskilled labor; capital, in this sense, meaning the savings from past production invested in tools and equipment to facilitate further production. Those intermediate products then become the raw materials for producing machinery and equipment for manufacturing, transportation, and communication, as well as for basic consumer products. Steel is the most obvious of these, though rubber and petrochemicals, as well as glass and copper refining, share some of the same characteristics.

In the interdependent modern world, not only the hope for higher wages and profits, but import substitution, pursuit of industrialization, or the need to achieve export capacity may be additional motives for a country to move to this stage. Newly industrializing countries such as South Korea and Brazil have their feet firmly on this step and have ambitions to move up to the next step.

To complete the analogy, the next way station is complex assembly, often relying on steel as the major intermediate product. Automobiles, shipbuilding, machine tools, household appliances, and consumer electronics characterize this stage. Capital-intensiveness is rising at this stage but so is the skill requirement. However, these new industrial workers have been accustomed to the earnings of less-advanced stages and for a time are prepared to give their economies the competitive edge of lower labor costs.

No one but the United States, western Europe, and Japan have matured enough to probe beyond the complex assembly stage. The emerging stage, the threshold of which these advanced nations have crossed, appears to include exotic processes and products such as precision castings and genetic engineering to produce new intermediate products such as specialty steels and chemicals, fine ceramics, and fiber optic cable, leading to ever more sophisticated lasers, large scale integrated circuits, advanced aircraft engines, artificial human organs, and so forth. Who knows what is next? What is critical at this point seems to be the sophistication of human skills and the responsiveness of technology without reliance on mass production. As this current top step on the industrial staircase emerges from the mists, there will undoubtedly be other steps beyond it.

Each stage is built upon those below which remain as essential supports. Any given country can either continue its earlier stages of production while launching into new ones or it can abandon the lower steps to those ascending the staircase below, depending upon international trade to provide products from the lower steps. It was within this framework that the American steel industry confronted international competition for the first time in the 1960s and beyond.

## THE ROLE OF STEEL IN ECONOMIC DEVELOPMENT

Steel is the epitome of the intermediate product-processing tread on the industrial staircase. It is a critical component of most of the intermediate and final products of an industrial economy, and is the prime symbol of entry into an industrial economy. Both the machinery for production and the products of an industrial society are made primarily of steel. Comparative advantage is the prime rule of economic life in an international setting. Each country should produce materials which it is best at producing and trade with other countries for those materials which it cannot produce at equal quality and cost. However, the country that would industrialize without its own steel capacity must import large amounts of machinery made of steel as well as semi-finished and finished steel as raw material for its final products. Generally, vast exports of extractive food, raw materials, or labor-intensive manufactured products will be required to pay for the import of steel. If import substitution—producing those items previously imported—is the preferred policy, steel production capacity will usually be among the first targets. With that first basic integrated steel mill also comes the badge of membership in the industrial society.

Steel production is characterized by economies of scale. For shoe and clothing manufacturing, the domestic demand is widespread and the production units are small and can be tailored to the income of the domestic market until foreign markets develop. Steel production is more capital-intensive, requiring larger amounts of capital per unit of production, than the simple consumer products manufacturing that precedes it, though less so than the complex assembly stage that follows it on the industrial staircase. Its average skill and wage level also lies between the stages above and below it.

Efficient steel production must begin with large output. The United States had both the population and the income distribution to intially absorb the output of its emerging steel industry.

So did Europe, but only as a total economic entity and not a political one. Brazil and India had the population potential but not the income distribution to absorb an efficient level of output. Therefore, they pursued steel production as part of a policy of import substitution. For countries like Japan, South Korea, and Taiwan, export was dictated by economy of scale pressures as well as by import substitution policies.

These factors led to a drive toward worldwide overcapacity during the 1970s and 1980s. This does not mean that the world was producing more steel than it needed, only more than could be purchased given the existing worldwide income distribution. Mass production requires mass markets which require mass purchasing power. The United States and western Europe accomplished internally the income distribution to make that possible. No such external distribution of income has yet been proven possible for the world at large.

Each stage of development is built upon those previous stages that remain as essential supports. A country either can continue its earlier stages of production, competing against newcomers, or it can abandon those stages to those on the previous step while depending upon trade to provide products from the lower steps. The focus at this point is on the confrontation that occurred as Pacific Rim newcomers challenged the steelmaking hegemony of North America and western Europe, joint venturing emerging as one means of accomodating economic conflict.

It should also be considered that our imagery of nations climbing an industrial staircase can be misleading. Though governments can establish policies which may accelerate or impede the economic development process, it is often enterprising individuals or groups of individuals organized into firms who identify the opportunities and undertake the risks. Succeeding or failing, they thereby provide the impetus which lifts their national economies to new heights, not challenging nations but challenging competing firms on the steps above them.

Starting from a knowledge of the historical development of iron and steel production in Europe and the United States, it is worthwhile to trace the development of steelmaking in Japan, Korea, and Brazil as well. That history illustrates a process now under way in Taiwan, Indonesia, and China which is likely to be followed later by others just now setting foot on the industrial staircase, all of whom are potential competitors to the present incumbents of the world steel industry. It was in response to those pressures that the IJVs which are the subject of this treatise emerged.

## THE WORLD'S STEELMAKERS

Developing the steel industries of Europe and the United States required more than a century because the technology and the skills had to be invented and developed by trial and error. Now this knowledge is available for the borrowing or the buying.

### The Industrial Pioneers

*Western Europe.* Great Britain was the first to place a foot on the industrial staircase, followed by other countries of western Europe. Starting with texiles, apparel, and shoes at the end of the eighteenth century, a railroad building boom in the 1830s demanded iron for rails, as well as for production machinery, and for the steam engines essential in the coal mines.[2] The introduction of the Bessemer converter in the 1850s and the open hearth furnace in the 1860s made it possible to transform ironmaking into steelmaking and, consequently, to create lasting equipment and machine tools. Those, in turn, were instrumental to a machinery industry, a mechanized agriculture, and so on.

In a neighborhood of small nations, it was not necessary for every country to grow equally through every stage. Also, no one country had the internal market to offer economies of scale for a full-scale steel industry until industrialization was well advanced. Therefore, though most of western Europe followed Great Britain in the basic consumer product stage, it was Germany that next launched into steelmaking and heavy machinery, passing Great Britain in steel production during the 1890s. Sixty years of production had dissipated the latter nation's comparative advantage in natural resources, though British labor and raw material costs were still low. An inefficient industrial structure accompanied by lagging technology was even more telling in the stagnation of the British steel industry which was demonstrated by the decrease of steel exports and the increase in imports.

By the 1860s, the Germany steel industry had introduced more advanced technologies than Britain while the ready market provided by the expansion of European railroads offset the advantage of lower British wage and raw materials costs during the 1870s and the 1880s.[3] After the 1890s, the German steel industry increased its competitive power by adopting the basic Bessemer process which enabled it to utilize cheap ores. The scale of plants was then enlarged and finally the continuous rolling process was introduced. The continued economic development of

Europe broadened markets until France, Belgium, and Luxemburg could also engage in steel production, all of which established a base for various countries to begin manufacturing machine tools and other goods. As a result, western Europe in aggregate was the world's largest steel producer and exporter until its productive capacity was dessimated by World War II.

World War II destroyed the industry of continental Europe and exhausted that of the United Kingdom. However, it left behind the necessary human resources—knowledge and skills—to revive industry as soon as the United States supplied the necessary capital for replacement of the technology. Western Europe imported steel from the United States for the first five years after the war and completed the reconstruction of its steel industries by 1950. However, the resurgent European industry was not a long-term competitive threat to the United States because its potential consumption capacity was adequate to absorb its output. European capacity continued rapid expansion on through the world steel boom of 1973–1974, drawing to a sudden close when that boom ended in a bust, driving home the message that world steel capacity was now beyond available demand with survival being a matter of both cost and quality.

*The United States.* The industrial development of the United States was at first impeded by British colonial policy but began with the American Revolution.[4] Textiles became a factory product early in the nineteenth century and stoves and shoes became products during the mid-century. The first blast furnaces and iron works date from 1765, but the steel industry in the United States also awaited the Bessemer converter which emerged in the 1860s. The first railroads in the United States also appeared in the 1830s but the great age of railroad building followed the Civil War and the availability of steel. With railroads, the machinery industry, agriculture, metal containers, oil and gas producers, and shipbuilding providing rising demand, the American steel industry grew dramatically during the last third of the nineteenth century.

The American steel industry caught up with British production by 1890, and during the first decade of the twentieth century surpassed Germany as the world's foremost steel producer.[5] The main sources of the American competitive power were the supply of low-cost ores and the introduction of hard-driving technology. Wages were higher and labor productivity was initially lower in America than in Europe. However, technology soon remedied the latter deficiency and by the beginning of the present century, the rising labor productivity in the U.S. steel industry more than made up for the high wages. However, all that would have been unlikely if the United States had not provided a massive domestic market to offer full range for the economies of scale. In 1913, the American steel industry produced more than 31 million tons, 42 percent of the world's total output.

The economies of scale in steel production did not require a continent-wide market; transportation costs were too high in relation to product prices for that to be an issue. Instead, a network of regional markets developed, with the high capital costs involved dictating a low number of firms in each of those markets. Beginning in western Pennsylvania and spreading across Ohio and Indiana to the Chicago area by World War I and then into the American South during the 1920s, the American West awaited war with Japan to create the need for substantial steel production.

The dawn of the postwar era brought with it a huge demand for steel to make up for the postponed consumption of the 1930s and 1940s. The industry resisted expansion, remembering the excess expansion during World War I and the long struggle with overcapacity which made the 1920s as well as the 1930s a veritable depression for the steel industry. Nevertheless, before the combined pressures of government policy and consumer demand, crude steel production increased from 66.9 million tons in 1940 to 96.8 million by 1950, and then reached 117 million tons in 1955. Bending with the experience, the industry then began to perceive its demand and growth to be limitless.

The period from the end of World War II in 1945 to the oil crisis of 1972–1973 became what was exaggerated as the American Century. It was a time in which the United States would adopt policies and practices perceived necessary to achieve world superiority. In economics, politics, and the military, the United States led the world. However, for at least the American steel industry, the policies of the period contained the seeds of their own destruction.[6] Too slow to invest at first, it suddenly did so in the 1950s during a burst of expansion in traditional open-hearth technology. But this only locked the U.S. firms into the technology of the past while European and, subsequently, Pacific Rim competitors were investing in new technology stressing basic oxygen and electric furnaces and the continuous caster.[7] During the 1950s, because of increased demand for steel which was derived from the Korean War and the rapid growth of American domestic economy, the U.S. steel industry added 50 million tons, bringing its capacity to 148 million tons. The United States continued its position as the world's largest and most efficient steel producer in the world until the end of 1950s when the reconstruction in Europe and Japan began to pose strong competition. The subsequent history of the American steel industry, and the events which led it into international joint ventures, can best be told after noting the rise of competition from the developing nations of the world.

### Steelmaking beyond the Pioneers

The emergence of three new steel exporters is relevant to our story; namely, Japan, South Korea, and Brazil.

*Japan.* Japan's pre–World War II role as a producer of cheap and rudimentary consumer goods requires no description. It was the ambition for military strength that first thrust Japan into steelmaking in the late nineteenth century.[8] It was also a deliberately chosen isolation that limited Japan's progress. British technology was imported, but the British technicians were rejected leaving the Japanese to learn by themselves through years of expensive trial and error. A fascistic relationship between the government and private enterprise primarily involved the steel industry in pursuit of national militaristic purposes. The long struggle for self-sufficiency made World War II possible.

Upon losing the war, Japan lost not only much of its steelmaking capacity, but its only significant market for steel—the military. Japan's crude steel production had peaked at 7.65 million tons in 1943 and then fell to a mere 0.56 million tons in 1946. It was the Korean conflict beginning in 1950 that put the Japanese back in the steel business by feeding a new military market. Moving aggressively to reindustrialize with American help, the Japanese industry produced 4.84 million tons in 1950 and 9.4 million tons in 1955.[9]

However, the very success in that sheltered market demonstrated the country's initial weakness in open competition. Initial quality weaknesses were attacked by an aggressive policy of steel modernization undertaken in 1956. Long-term raw materials purchase contracts were concluded with existing suppliers in Australia and Africa and new mines were developed with Japanese financing in the Philippines and Malaysia. Steelmaking facilities were located adjacent to deep water ports, hauling the iron ore and coking coal in massive ships of their own building.

Prior to 1950, Yawata Steel, Fuji Steel, and Nihon Okan were Japan's only integrated mills. When Yawata and Fuji merged as

Nippon Steel in the early 1950s, the new company had one-tenth the capacity of the U.S. Steel Corporation, even though Nippon Steel was soon to become the world's largest steelmaker. Soon, non-integrated steel firms such as Sumitomo Metal Industries, Kawasaki Heavy Industries, and Kobe Steel Works integrated; they did so, along with other basic firms, with the most advanced basic oxygen furnaces and continuous casters. Between the 1950s and early 1970s, the Japanese steel industry added eleven new facilities which in 1980 accounted for 80 percent of total Japanese steel production.

What proved to be a more lasting problem was the fact that steel was needed as a base for other industries, yet there was not adequate consumer demand to support either a viable steel industry nor a domestic industrial market using steel as an intermediate product. Output had risen to 22.14 million tons by 1960 and peaked at 120 million tons per year by 1973. Despite the rapid growth of its auto and other industries, the Japanese industry could not begin to absorb that output. It was not import substitution but export development which was the primary objective.

By 1955, Japan was exporting 6.6 percent of the world's steel compared to 14.2 percent for the United States, 9.9 percent for West Germany, and 10.6 percent for the Soviet Union and eastern Europe. Japan's share of world steel production moved from 6.5 percent in 1960 to 15.6 percent in 1970. In addition, Japan's share of world exports of steel continued to rise to 16.1 percent in 1965 and 28.6 percent in 1973, becoming by far the world's largest exporting country.[10]

At first, Japanese steel suffered in productivity as well as quality but soon caught up in both. In 1958, a ton of hot-rolled Japanese sheet steel required three times as many man-hours in production as American steel. By 1964, the ratio was two to one, and in 1972 it was even. By 1980, Japanese man-hour productivity was ahead by a margin of 0.94, 5.36 to 4.42. There were also cheaper hours, paying $2.31 per hour in Japan compared to $12.50 per hour in the United States as of 1966, and rising to a Japanese rate of $10.89 compared to an American rate of $24.67 by 1982.

By the early 1960s, the Japanese steel industry had established itself as one of the world's most efficent industries, and had also become the world's largest exporter of steel by 1969. Within that massive increase in modernized production capacity was included experimentation with new alloys and other high-quality products to meet the demands of their rapidly growing automobile needs and other industrial customers. The Japanese firms also covered their bets by diversifying themselves into new businesses based upon related technologies such as steel by-products, computers, and environmental controls. Yet by the mid-1970s, Japan was beginning to experience decline in both output and employment because it failed to develop its domestic market and began to face competition from South Korea for its two most important export markets—the People's Republic of China and the United States. However, by then Japan had moved its industrial base into more complex products and had become a major exporter of steel technology, including exports to its neighbor and emerging competitor, South Korea. Further, and most relevant to our study, Japanese steel companies began looking for the partners for IJVs to secure their overseas markets.

*South Korea.* Until 1910 when it was taken over by Japan, the Korean steel industry was an insignificant size and was based on simple manual technology, making only a few consumer goods. Thereafter, Japan built several steel mills on the Korean peninsula, most of those mills still in production having been destroyed during the Korean War of 1950–1953. From then until 1973, the Korean steel industry consisted of four companies with small and outdated facilities.[11]

The presence of the United States following its conflict with North Korea was the primary factor in launching South Korea into the modern industrial world, with steelmaking being an essential element of that launch. South Korea lacked not only iron ore and coking coal but also the capital necessary for building a new integrated steel mill. Its own domestic market was too small to absorb the output of even one integrated mill of efficient size. The newly industrializing South Korea would be confronting international competitors with high technology and stable markets. In addition, its nearby market was the world's most efficient steelmaker—Japan.

Despite these disadvantages, the Korean government planners believed that a prosperous steel industry could provide the impetus necessary to maintain a progressive economic development. The domestic demand for steel would expand in concert with economic growth. The linkage effects with steel, which are always of critical importance in the context of a developing economy, would then, it was assumed, stimulate and transform the whole manufacturing sector.

Korean industrial planners attempted to launch an integrated iron and steel mill early in the 1960s. However, the World Bank rejected the loan request, arguing that because it lacked raw materials, Korea was a poor risk for steel investment. The Korean government persisted and finally succeeded in forming the, Pohang Steel Company (POSCO) in 1968 and persuading the Japanese government to provide finance from Japanese war reparations and substantial loans from the Japanese Export–Import Bank. The absence of democracy consequent to a military government was expected to help contain labor costs. Whether the Japanese realized they were creating a competitor or an economic colony, or were merely hungry for the construction and machinery contracts, is not clear. However, Korean industrial development has clearly been beneficial to the Japanese economy.

The first integrated steel mill in Korea began operation in 1973 and has operated at a profit since its beginning, in part because of substantial government support. POSCO has been supported by low-interest foreign loans guaranted by the Korean government, substantially discounted user rates for railroad services, and a small amount of social overhead capital. POSCO's high quality and low-cost steel have been important factors to assisting Korean producers of automobiles, construction, containers, and machinery in becoming competitive within the international market. During the early 1980s, the Korean government began to make plans for a second integrated steelmill and decided finally on Kwangyang as the site for the plant which would open in 1987. By the end of 1992, POSCO's total capacity had reached 21 million tons per year. A substantial proportion of that output, as we shall see in the next two chapters, was dependent upon POSCO's involvement in a joint venture with the American steelmaker USX.

*Brazil.* Well endowed with all the raw materials of steelmaking except for good quality metallurgical coal, Brazil has produced forgeable iron since colonial times. Brazil's movement into steel production was internally oriented from the start. Because of its forward and backward linkages, steel was viewed as essential in government strategy to stimulate vertical integration within the emerging industrial complex. The Brazilian government formally pursued an import substitution strategy as early as 1909 and sought expanded steel production by offering monopoly privileges and subsidies to attract foreign capital.[12]

Difficulties in importing steel during World War I strengthened the Brazilian push for steel producing capacity. In 1931 President Vargas stated, "The most basic problem of our economy is that of steel. For Brazil, the age of iron will signify its economic opulence." After this statement, Brazil began to court

American and German investors to underwrite the expansion.[13] By the early 1940s, the Brazilian government had parlayed increasing worldwide tensions into sizeable loans from the U.S. Export–Import Bank and formed a government-owned and government-operated steel company that by 1948 built the country's first integrated mill.

Formalized economic planning following World War II led Brazil into the formulation of a national steel policy primarily dedicated to import substitution. Beginning in the 1950s, the Brazilian steel industry began to draw heavily on Japanese technology and financing; and the country moved toward self-sufficiency in steel capacity.

Brazilian economic policy became more outward looking in the mid-1960s with a change in the political regime. Export taxes were abolished, administrative proceedures for exporters were simplified, tax incentives for exporters were instituted, and the inflow of foreign capitial accelerated sharply. Events of the early 1970s, such as the oil crisis and the accompanying need to increase foreign exchange earnings, were major factors pulling and pushing Brazilian steel production into an increased posture of export promotion while also motivating massive internal investment in steel, metal products, petrochemicals, and capital goods to stall any renewed import dependence for industrial growth.

Though charcoal-fired blast furnaces still domestically produced consumed iron, by the late 1970s, Brazil's government-owned steel works were among the world's most modern, with nearly half of total Brazilian steel producion relying on continuous casting. Building from that base, Brazil entered the world of heavy machine production and complex assembly such as aircraft. Still, domestic demand and manufacture for export was not adequate enough to provide the desired market for Brazilian iron ore and steel, which also led to the joint ventures described in subsequent chapters.

## Other Challengers

Other countries are in line waiting to climb the industrial staircase, including steel exportation. Only European mills and the three newly industrializing nations cited have thus far been significant challengers to the brief U.S. dominance in steel production. Looking to the future, Taiwan and the Republic of China have the essential industrial discipline to become significant challengers. Indonesia and the Philippines have access to both the necessary raw materials and human resources but are plagued by internal corruption. India and Mexico are examples of nations that have significant steel capacity but place higher priority on job creation than on productive efficiency; and, until recently, they have been hiding behind a screen of trade protection. India, for instance, had a level of steel production just below that of South Korea in 1983 and 1984 but employed over 300,000 steelworkers compared to the latter's 70,000.[14]

The most significant fact for the 1970s and 1980s was that every steelmaking country outside the communist bloc was increasing its production capacity more rapidly than it was increasing its consumption of steel (Table 3.1).

In fact, all but South Korea and Brazil were actually reducing their consumption and, at the same time, expanding their capacity. Since a few other less-developed countries such as Mexico and India were self-sufficient in steel and no non-steel producing country other than the Middle East oil producers had the necessary purchasing power to consume the steel they needed, world steel production capacity far exceeded potential consumption. Further, as the world's richest nation, the United States was the preferred target for every country's marketing efforts, even though its own capacity exceeded its own consumption.

**Table 3.1**
**Capacity versus Consumption for Major Countries (millions of metric tons)**

| | Crude Steel Capacity | | Percent Annual | Apparent Steel Consumption | | Percent Annual |
|---|---|---|---|---|---|---|
| | 1969 | 1983 | Growth | 1969 | 1983 | Growth |
| Belgium/ Luxembourg | 19.9 | 27.9 | 2.8 | 4.6 | 2.9 | -2.8 |
| France | 25.2 | 32.1 | 1.9 | 23.0 | 15.0 | -2.5 |
| WestGermany | 49.7 | 66.2 | 2.4 | 40.9 | 29.8 | -1.9 |
| Italy | 19.5 | 42.7 | 8.4 | 20.2 | 18.1 | -0.7 |
| United Kingdom | 29.1 | 26.7 | -0.5 | 24.9 | 14.0 | -3.1 |
| Japan | 81.0 | 175.6 | 8.3 | 67.1 | 65.6 | -0.2 |
| United States | 140.5 | 149.9 | 0.5 | 132.9 | 94.5 | -2.1 |
| South Korea | 0.7 | 13.5 | 130.6 | 1.3 | 8.9 | 41.7 |
| Brazil | 4.6 | 22.2 | 27.3 | 5.5 | 8.1 | 3.4 |

*Sources:* 1969 data, U.N. statistics for capacity, OECD statistics for apparent consumption; 1969 data for Korea and Brazil, Korean Iron and Steel Federation, *Statistical Yearbook, 1976;* Consumption for 1983, *Statistical Yearbook, 1986;* Capacity for 1983, Chase Econometrics, *The World Steel Outlook to 1995,* vol. 1, July 1983.

## BACK TO U.S. DEVELOPMENTS

The steel industries of those countries competing with the United States have been either government-owned, heavily government-influenced, or, which is often the case, government-subsidized. The private and unsubsidized American industry could either compete by achieving lower costs and higher quality, by seeking to gain protection from its own government at high cost to American consumers and taxpayers, or by withdrawing from the competition—it would attempt all three to varying degrees. But long before that intense world competition had become apparent, the U.S. industry had already handicapped itself by several strategic mistakes.

### Industry Expansion

Had the American steel industry perceived itself in the 1960s as standing with the western Europeans on the basic industry step of the industrial staircase with a pack of newly industrializing countries panting to displace them, then the U.S. companies would probably have followed a much different investment pattern. As noted earlier, U.S. steelmakers were reluctant to expand at the end of World War II, fearing the overcapacity which overcame them following World War I. By the mid-1950s, with continued high consumer demand at home, the Cold War, the Korean conflict abroad, and little foreign competition yet apparent, steel marketeers moved to an opposite extreme. They now acted as if they believed that high output would last forever interrupted only temporarily by the recurrent business cycle.

U.S. industry leaders had not invented these favorable forecasts by themselves. They were constantly prodded to expansion by government and academic advisers. As late as 1973 and 1974, the International Iron and Steel Institute was convinced that capacity was and would remain the only constraint on output, while the American Iron and Steel Institute, in 1975, forecast a U.S. capacity need of 170 million tons by 1983.[15] Actual production in the latter year turned out to be 85 million tons with the 151 million tons of 1973 having been the peak of U.S. production. Consumption was down as the nation pulled out of the deep 1981–1982 recession, but not by much. What could not be known at the time was that 1974's 781 million tons would end growth in world steel production, a figure which over the next twenty years would be equaled only once in 1989.[16] Steel consumption had stagnated in the United States in part because of the shift to substitute materials. The difference between 1969 and 1983 consumption could be largely accounted for by the steel in imported automobiles and other products.

After resisting government proposals for industry expansion during the last half of the 1940s, the American industry expanded its capacity aggressively from 100 to 148 million tons during the decade of the 1950s. It did so, however, almost totally by replicating existing technology, right at the time that its future competitors were rebuilding their war-destroyed industries with exotic new developments. Capacity was increased, but not productivity.

The U.S. industry increased its traditional open-hearth capacity from 90 million to 126 million tons during the 1950s, bypassing the more efficient basic oxygen furnace (BOF), which was invented by the Austrians in 1952, to perform the same task of transforming pig iron into steel. During the early part of the decade, the BOF was considered experimental and had been applied only to small operations. By the end of the decade, this was the world standard adopted by the Germans and Japanese as they rebuilt and by the South Koreans as they entered steel production for the first time. The difference was the ability to melt a 300-ton heat in forty-five minutes rather than six hours, especially important as energy costs multiplied during the 1970s. At the same time, North American ore reserves were expanded, even though U.S. firms had access to richer and cheaper South American, African, and Australian ore reserves.[17]

By 1960, U.S. producers recognized their technological misstep, but it was too late. The industry was already in trouble and did not have ready access to the capital needed to catch up. With imports increasing, the domestic industry was confronted by overcapacity. Falling profits reduced available retained earnings and made steel unpopular in the stock market. Added capitalization would have to come from debt which the industry had limited ability to amortize. Table 3.2 traces the international path of technological innovation throughout this period.

One anomaly of the development of surplus capacity by the Europeans and the Japanese was that they attempted to export when their own demand was slack and they withdrew to meet their own domestic needs when the demand was high. Therefore, they tended to amplify the U.S. business cycle. Import penetration rose during the years of U.S. prosperity in the 1960s but decreased during the late 1960s and early 1970s as a consequence of a worldwide industrial boom. Hereafter, the U.S. industry was

**Table 3.2**
**Adoption of New Technologies**

| | USA | | Japan | | EEC | | Canada | |
|---|---|---|---|---|---|---|---|---|
| | Percent | Metric tons | Percent | Metric tons | Percent | Metric tons | Percent | Metric tons |
| Basic oxygen furnace | | | | | | | | |
| 1960 | 3.4 | 3.3 | 11.9 | 2.9 | 1.6 | 1.8 | 28.1 | 1.6 |
| 1965 | 17.4 | 22.9 | 55.0 | 24.9 | 19.4 | 24.3 | 32.3 | 3.3 |
| 1970 | 48.1 | 63.3 | 79.1 | 81.2 | 42.9 | 65.1 | 31.1 | 3.8 |
| 1975 | 61.6 | 71.8 | 82.5 | 92.9 | 63.3 | 87.2 | 56.1 | 8.0 |
| 1981 | 60.6 | 73.2 | 75.2 | 84.1 | 75.1 | 103.6 | 58.6 | 9.4 |
| Basic oxygen furnace plus electric furnace | | | | | | | | |
| 1960 | 11.8 | 11.7 | 32.0 | 7.1 | 11.5 | 12.4 | 40.4 | 2.3 |
| 1965 | 27.9 | 36.7 | 75.3 | 34.1 | 31.5 | 39.5 | 45.1 | 4.6 |
| 1970 | 63.5 | 83.5 | 95.9 | 98.4 | 57.7 | 87.6 | 45.9 | 5.6 |
| 1975 | 81.0 | 94.5 | 98.9 | 111.3 | 82.6 | 113.7 | 76.4 | 10.9 |
| 1981 | 88.8 | 107.3 | 100.0 | 111.9 | 98.6 | 136.0 | 86.5 | 13.9 |
| Continuous casting | | | | | | | | |
| 1971 | 4.8 | 5.8 | 11.2 | 11.0 | 4.8 | 6.7 | 11.5 | 1.4 |
| 1976 | 10.5 | 13.5 | 35.0 | 41.4 | 20.1 | 29.7 | 12.0 | 1.7 |
| 1981 | 21.1 | 25.3 | 70.7 | 79.0 | 45.1 | 62.3 | 32.2 | 5.3 |

*Source:* Donald F. Barnett and Louis Schorsh, *Steel: Upheaval in a Basic Industry* (Cambridge, Mass.: Ballinger, 1983), 55.

tempted to expand again, this time by resorting to basic oxygen furnaces, only to have the oil price multiplication by the Organization of Petroleum Exporting Countries (OPEC) dampen world growth and generate the deep 1974–1975 U.S. recession.

This time there resulted an increase in productivity and capacity as well as an increase in debt service costs. Imports came flooding back in the mid-1970s, since the OPEC-generated recession hurt other world steel producers as well; capacity utilization dropped sharply and profits followed.

A substantial part of the U.S. capacity expansion occurred completely outside the traditional steel industry with the emergence of the upstart mini-mills.[18] They were called mini-mills, not necessarily because of their size but because they bypassed the coke ovens and blast furnaces used to make pig iron out of iron ore. They limited themselves to melting scrap in electric furnaces and pouring the molten metal directly onto continuous casters to turn out such "low end" products as reinforcing rod, angle iron, and I beam, primarily for the construction industry. Since this industry was non-union, they paid low hourly rates supplemented by generous production-determined bonuses, utilizing only about two hours of labor per ton of steel compared to the six to eight hours the integrated mills required well into the 1980s. Therefore, the mini-mills had cost structures competitive with the international competition. By 1984, they were producing 18 percent of total U.S. steel output.

In summary, at a time during the 1950s when, in retrospect, the basic industry should have concentrated on cost-cutting investments and the abandonment of outmoded capacity, it had again expanded along more traditional lines. That the investment was not rewarded by profits seems to have been disillusioning enough to have initiated a diversification binge. The result for the basic steel companies was a pronounced trend toward disinvestment during the 1980s.

### Oligopoly Pricing

The American industry also followed an oligoply pricing policy, acting as if it would continue to have no competitors. When there are few producers of an essentially homogeneous product (can the purchaser of an automobile tell whether the steel in it came from U.S. Steel or Bethlehem, and would they care if they could even though they may care greatly whether the automobile is a Ford or a Chevrolet?), no one producer can change prices without considering the reactions of each of the others. If one firm cuts prices in order to win sales from its competitors, the other producers follow suit to preserve customers; consequently, all are worse off than they were previously. If one firm raises prices and the others are not willing to follow, the initiator must reverse the decision or be abandoned by all purchasers. Hence, price competition had given way earlier to price leadership, usually by the U.S. Steel Corporation which only changed its prices if confident that the others would follow. The timing of such an increase might be sparked by some production cost increase common to all steel producers, but more often the increase would follow a recognized general increase in the demand for steel which would support desired production levels at a higher rate of profit. On the other hand, in a recession, rather than cut prices to maintain demand, the tendency was to maintain price and let the fluctuation occur in production even though there were often "under the table" concessions to purchasers in terms of extras in quality and service. Taking the fluctuation in output rather than price was amplified in its impact on profits by the fact that an industry of high fixed costs, such as steel, tends to have a relatively high break-even point. Hence, the industry's profit fluctuations were even wider than its output fluctuations.

The most common pricing component among competitors was labor costs. In 1937, Myron Taylor, U.S. Steel chairman, had even sought and signed a labor agreement with John L. Lewis's Steel Workers Organizing Committee during the gestation of the United Steel Workers of America, hoping that a uniform wage would help stem the depression drive toward price cutting. This strategy didn't work because most of the other companies resisted unionization until pressured by wartime labor shortages. With the end of the war's wage and price controls, a tacit agreement between the industry and the union emerged. The union's demands for wage increases would be immediately passed on in price increases with a minimum of production interruption. Until 1956, the wage settlement process took the form of pattern bargaining, and U.S. Steel would settle with the steelworkers first. This occurrence was often based on the level of increases the United Auto Workers had gotten by pattern bargaining among the steel industry's single most important set of customers. U.S. Steel would be reluctant to take a strike while its competitors produced steel. Then, after U.S. Steel had granted a favorable settlement, the union would demand it of the others who would be reluctant to accept a shutdown while U.S. Steel and other firms supplied steel customers. Further, since all of these firms were experiencing the same labor costs, none of them were hurting in the short run as long as there were no steel suppliers outside the circle. However, this era ended with foreign competition as it arose during the 1960s.

In 1956, the industry ended the pattern bargaining pressure by insisting upon industrywide bargaining. All fifteen major producers met simultaneously with steelworker leaders. The chief bargainer for U.S. Steel spoke for all of the companies, accompanied at the table by the representative of Bethlehem or, occasionally, one of the other larger firms. The major economic conditions bargained for by the U.S. Steel representative (with his second's acquiescence) was then accepted by the entire industry and spread from there to its satellite suppliers and fabricators. Special conditions, applicable to only one firm, might then be negotiated in side agreements before each signed identical copies of the master agreement. All of this worked well in an oligopolistic structure but could not be maintained as long as there were significant producers outside of the circle which, in this case, meant foreign ones. With the emergence of international competition, the consequences were dire.

### Abandonment and Restoration of Wage Restraint

Government played an unintentional but critical role in the crisis which emerged. In formulating negotiation strategies, union negotiators must estimate a given management's willingness to make concessions. The extended length of the strike is likely to be necessary to win any specified level of concessions as well as the willingness of members to accept that level of current income loss in anticipation of long-term gains. Management bargainers make simultaneous estimates of the union's staying power in forming their strategies. Convinced that with both cold and hot war the government would not allow the entire steel industry to be shut down, the steelworkers raised their demands during the late 1940s and 1950s above what they might otherwise have sought. They did so knowing that the government would intervene in any stoppage, pressuring management to make concessions to settle the dispute. Knowing the game, management lowered its offers to save some "caving-in money" to give up under government pressure and at the same time demanding government acceptance of compensating price increases, perhaps with an additional surreptitious bonus. It was a formula almost guaranteed to produce many short strikes and accelerating wages and prices.[19]

That act in this drama came to an end in 1959 when Secretary of Labor James P. Mitchell and Vice President Nixon convinced President Eisenhower not to intervene. The strike went on for 116 days after which the parties essentially settled for what management would have granted a few days into the strike. In the meantime, steel users did not sit idly by waiting for supplies: They went looking for new suppliers and found them in the emerging European and, later, Japanese mills. For the first time ever, imports became an important factor in American steel supplies, as Table 3.3 indicates.

The 1959, 116-day strike hurt everyone involved so badly that it became the last industrywide strike in the steel industry, but

**Table 3.3**
**U.S. Trade in Steel Mill Products (in thousands of net tons)**

| | Production | Exports | Imports | Consumption | Import Penetration (Percent) |
|---|---|---|---|---|---|
| 1940 | 45,966 | 7,640 | 18 | 38,344 | 0.0 |
| 1945 | 57,242 | 4,354 | 54 | 52,942 | 0.1 |
| 1950 | 72,232 | 2,639 | 1,014 | 70,607 | 1.4 |
| 1960 | 71,149 | 3,359 | 3,359 | 71,531 | 4.7 |
| 1965 | 92,666 | 2,496 | 10,383 | 100,553 | 10.3 |
| 1970 | 90,798 | 7,062 | 13,364 | 97,100 | 13.8 |
| 1971 | 87,038 | 2,827 | 18,304 | 102,515 | 17.9 |
| 1972 | 91,805 | 2,873 | 17,681 | 106,613 | 16.6 |
| 1973 | 111,430 | 4,052 | 15,150 | 122,528 | 12.4 |
| 1974 | 109,472 | 5,833 | 15,970 | 119,609 | 13.4 |
| 1975 | 79,957 | 2,953 | 12,012 | 89,016 | 13.5 |
| 1976 | 89,447 | 2,654 | 14,285 | 101,078 | 14.1 |
| 1977 | 91,147 | 2,003 | 19,307 | 108,451 | 17.8 |
| 1978 | 97,935 | 2,422 | 21,135 | 116,648 | 18.1 |
| 1979 | 100,262 | 2,818 | 17,518 | 114,962 | 15.2 |
| 1980 | 83,853 | 4,101 | 15,495 | 95,247 | 16.2 |
| 1981 | 88,450 | 2,904 | 19,898 | 105,444 | 18.2 |
| 1982 | 61,567 | 1,842 | 16,663 | 76,388 | 21.8 |
| 1983 | 67,584 | 1,199 | 17,070 | 83,455 | 20.5 |
| 1984 | 73,739 | 980 | 26,163 | 98,922 | 26.4 |
| 1985 | 74,000 | 900 | 23,000 | 96,100 | 23.9 |

*Source:* American Iron and Steel Institute, *Annual Statistical Report*, various years. Import and consumption figures do not include the steel content of imported automobiles, machinery, etc.

that didn't matter. Once bitten, steel users stockpiled in anticipation of each subsequent labor negotiation. Then, when no strike resulted, they canceled orders until they used up their excess inventories, requiring layoffs and idling of capacity as devastating to the industry as if there had been a strike.

Apparently, seeing the threat of being perceived as an unreliable supplier to be more damaging than becoming a high-cost supplier, the steel employers, after a few interim steps in that direction, proposed and gained acceptance in 1973 of what it called an Experimental Negotiating Agreement (ENA). The ENA consisted of an automatic 3 percent annual wage increase which was adapted to the annual productivity increases of the entire economy rather than the much lower industry productivity trends and further buttressed by an uncapped cost of living allowance (COLA). Strikes were forbidden, uninterrupted production guaranteed, and subsequent disputes would be subjected to binding arbitration.

Inflation during the Vietnam conflict had increased the union's demand for a COLA, yet the U.S. withdrawal from that conflict had reassured the industrywide bargainers that the COLA would not be expensive. As it turned out, this was wrong. Since 1973 was also the year of the OPEC oil price revolution with all of its inflationary consequences, the COLA generated spectacular wage increases which, along with the automatic increases, were then added permanently to the basic wage level as the base upon which all employee benefits were calculated.[20] Steel wages spiraled upward without further resort to negotiation from 1.4 times the average manufacturing wage, where they had stood in 1970, to nearly double the average by 1982 (see Table 3.4). Though generally considered to be capital intensive in its production processes, an industry in which 30 percent of total costs were labor costs could not blithely ignore the consequences.

The ENA guaranteed labor peace, but it also eliminated the opportunity for any strike-threatening or strike-taking confrontation to end the upward wage spiral. As the weaker firms in the steel industry were threatened by closures, the Steel Workers Union granted concessions to save its members' jobs which antagonized the larger firms in the process. Since overcapacity was rife and downsizing was essential, saving the weaker firms meant that the union was guaranteeing that the stronger firms would have to shrink more than they otherwise might have done.

U.S. wages were already higher than those of any competing nation (Table 3.5) and were sustained by its superior productivity (Table 3.6). However, both productivity rates and wage costs were rising more rapidly in the competing nations causing the U.S. advantage to shrink.

After 1980, the industry employers exercised their right to cancel the ENA approach. By 1983, inflation would have largely run its course, but no one knew that at the time. The employers came to the 1983 negotiations committed to a wage rollback and facing a union which was more concerned with preserving jobs

**Table 3.4**
**Hourly Wage Trends in the U.S. Steel Industry and All Manufacturing (U.S. dollars)**

| Year | Steel Industry | All Manufacturing | Index for Steel (all mfg.=100) |
|---|---|---|---|
| 1935 | .655 | .544 | 120 |
| 1940 | .855 | .655 | 131 |
| 1945 | 1.257 | 1.016 | 124 |
| 1950 | 1.746 | 1.440 | 121 |
| 1955 | 2.509 | 1.860 | 135 |
| 1960 | 3.349 | 2.260 | 148 |
| 1965 | 3.935 | 2.610 | 151 |
| 1970 | 4.767 | 3.350 | 142 |
| 1973 | 6.331 | 4.090 | 154 |
| 1975 | 8.311 | 4.830 | 172 |
| 1980 | 14.177 | 7.270 | 195 |
| 1981 | 15.449 | 7.990 | 193 |
| 1982 | 16.808 | 8.490 | 198 |
| 1983 | 16.094 | 8.830 | 171 |
| 1984 | 15.343 | 9.180 | 167 |

*Sources:* For steel industry: American Iron and Steel Institute, *Annual Statistical Report*. For all manufacturing: U.S. Department of Labor, Bureau of Labor Statistics.

and protecting laid-off workers than in further wage gains given the rapid decline in employment of its members (Table 3.7). A wage cut of approximately 10 percent, which was promised to be gradually restored in the following three years, was negotiated. Holiday and vacation concessions were made and the COLA was restricted, but the supplementary unemployment benefits were strengthened. The cuts were modest under the circumstances. Work rules relaxation may have been even more important in the long run.

In anticipation of the 1986 negotiations, steel management announced abandonment of the industrywide bargaining practice which had prevailed since 1955. Each company would pursue

**Table 3.5**
**Steel Industry Hourly Employment Costs (U.S. dollars per hour at current exchange rates)**

| Country | 1966 | 1982 | Percent increase | 1985 | Percent increase |
|---|---|---|---|---|---|
| Japan | 2.31 | 10.89 | 371 | 11.00 | 1 |
| West Germany | 4.70 | 13.27 | 182 | 10.00 | -25 |
| France | 4.53 | 12.14 | 167 | 10.00 | -18 |
| United States | 12.50 | 24.67 | 97 | 23.00 | -7 |
| United Kingdom | 4.12 | 9.14 | 121 | 5.50 | -40 |
| South Korea | 0.30 | 1.84 | 513 | 3.00 | 68 |
| Brazil | n.a. | 4.00 | n.a. | n.a | n.a. |

*Sources:* For the United States, Japan, West Germany, France, and the United Kingdom: Paine-Webber, *World Steel Dynamics, Steel Strategist #9,* February 1984; *World Steel Dynamics,* March 11, 1985; for South Korea: South Korea Iron and Steel Federation, *Statistical Yearbook, 1982* (Seoul), pp. 301–331; *World Steel Dynamics,* March 11, 1986; for Brazil: Hans Mueller, "Trends in Steel Production and Trade," paper no. 92 delivered at the Annual Conference of the Eastern Economic Association, New York, March 14–18, 1984, p. 41.

**Table 3.6**
**Labor Productivity for All Steel Industry Employees (hours per net ton at actual operating rates)**

| Country | 1966 | 1982 | Percent improvement 1966-82 | 1985 | Percent improvement 1982-85 |
|---|---|---|---|---|---|
| Japan | 22.4 | 8.0 | 64 | 7.2 | 11 |
| W.Germany | 22.5 | 11.1 | 51 | 9.5 | 14 |
| France | 23.8 | 10.8 | 51 | n.a. | n.a. |
| U.S. | 12.1 | 7.8 | 35 | 6.5 | 17 |
| U.K. | 25.5 | 13.4 | 48 | n.a. | n.a. |
| S.Korea | n.a. | 15.0 | n.a | 14.2 | 6 |
| Brazil | n.a. | 27.9 | n.a. | n.a. | n.a. |

*Source:* American Iron and Steel Institute, *Annual Statistical Report,* various years.

**Table 3.7**
**Employment in the U.S. Steel Industry from 1973–1985**

| Year | Industry Employment | Percent Change |
|---|---|---|
| 1973 | 604,600 | +6.4 |
| 1974 | 609,500 | +0.8 |
| 1975 | 548,200 | -10.1 |
| 1976 | 549,400 | +0.2 |
| 1977 | 554,300 | +0.9 |
| 1978 | 560,500 | +1.1 |
| 1979 | 570,500 | +1.8 |
| 1980 | 511,900 | -10.3 |
| 1981 | 506,100 | -1.1 |
| 1982 | 394,300 | -22.1 |
| 1983 | 343,100 | -14.7 |
| 1984 | 334,000 | -2.7 |
| 1985 | 237,500 | -28.9 |

*Source:* Bureau of Labor Statistics.

its own self-interest in negotiations with its own employees. Preliminary negotiations suggested the possibility of major departures. Weirton Steel sold out to its employees through an employee stock ownership program (ESOP). Wheeling–Pittsburgh Steel, from behind the barricade of a Chapter 11 bankruptcy proceeding, negotiated a $5-per-hour cut in labor costs. The negotiation was won only at the price of the removal of the company president and senior management and initiation of joint worker–supervisor committees to oversee operations. LTV Steel Corporation traded profit-sharing and preferred stock for a $3.60-per-hour cut in labor costs. National Steel persuaded its employees to accept a move in the direction of the compensation practices of its Japanese partner, Nippon Kokan. In exchange for profit-sharing, incentive pay tied to productivity, and a company promise to spend $1.2 billion on modernization, the employees accepted a 6.4 percent reduction in labor costs. But, most important, in return for a relatively free hand in manpower and work rules, the company promised job security with employment reduced only by attrition and sweetened early retirement which was expected to save $100 million per year in the process.

### U.S. Government Policy

As if all of that was not bad enough, the U.S. government policies of the period worsened the conditions produced by faulty management and union policies. During 1945–1959, the federal government applied pressure for compromise to prevent or end steel strikes which contributed to an additional upward bias in wage settlements. This cheered on the initators of the ENA, heralding it as a "historic step" emerging from "courage and creativeness." Government pressures to restrain steel prices during the war in Vietnam discouraged generation of cash flows that could have been used for modernization but might also simply have been used for dividends and diversification. Tax policies provided greater rewards to mergers, diversifications, and real estate scams than to investment in the modernization of basic industries.

In response to industry pressure in the 1970s and again in the 1980s, the federal government restricted imports through various protectionist devices which reduced pressure in order to make any needed productivity-improving or cost-reducing adjustments. Swollen federal budget deficits and high interest rate policies in the 1980s resulted in a stronger dollar in relation to other currencies and created an exchange rate situation that lowered the price of imports by 40 percent while raising the price of U.S. exports by 70 percent in world markets (see Table 3.8). With the 1984 exchange rates, the United States was the world's highest-cost steel producer, whereas, had the exchange rates of the late 1970s prevailed, only Japan would have had lower production costs and not by a substantial margin.

**Table 3.8**
**Production Cost Consequences of Exchange Rates (costs per net ton at actual operating rates)**

| Country | At 1984 Exchange Rates | At 1978-79 Exchange Rates | Rate Distortion |
|---|---|---|---|
| U.S. | 478.06 | 478.06 | - |
| Japan | 447.66 | 454.05 | 1.4 |
| W.Germn | 416.63 | 481.54 | 15.6 |
| France | 423.11 | 645.11 | 52.5 |
| U.K. | 397.33 | 489.52 | 23.2 |
| S.Korea | 297.86 | 493.68 | 65.7 |

*Sources:* Paine-Webber, *World Steel Dynamics, Steel Strategies #9*, February 1984, and authors' calculations.

U.S. producers complained that they faced something less than a level playing field in international competition. The steel industries of competing nations were largely government owned or government subsidized.

Even though existing legislation had forbidden foreign firms from selling their products in the United States below cost, or "dumping," Figure 3.2 illustrates the extent to which Japan was willing to subsidize its own steel exports during the period from 1973–1983 in order to keep its mills humming. To sell below total cost but above variable cost is rational economic action in any industry characterized by high fixed costs. Since labor was also, to some degree, a fixed cost to European employers with limits on their abilities to lay off unneeded employees, European producers had an even greater economic incentive to sell their products below total cost. Total European operating subsidies for the period between 1975–1985 were estimated at $30 billion at the exchange rates of the time, in addition to an estimated $34 billion in government capital investment.[21]

The unsubsidized U.S. industry operated in the black until the deep 1981–1982 recession and then fell into a loss position. All European mills except for those in West Germany lost money

**Figure 3.2**
**Real Import Prices and Full Japanese Costs for Carbon Steel from 1973–1983**

throughout the 1970s. Only the South Korean steel industry made a profit every year throughout this period, while the Japanese, not including government subsidies, experienced no losses until 1983 (Table 3.9).

New steel-producing capacity in the developing world was being financed in part by loans from U.S. banks, while at the same time those same banks were unwilling to loan to U.S. steel firms because of their low rates of return. Further, when Brazil was unable to meet the interest payments on its huge borrowings, the loaning U.S. banks opposed any trade restrictions which might prevent Brazilian steel from reaching the U.S. market.

Partly out of disillusionment with the results of the Smoot-Hawley Tariff of the 1930s which had contributed to the worldwide spread of depression, and partly because of the internationalist commitments with which it embraced the United Nations and related developments at the end of World War II, but mostly because it had the world's most productive industrial machine and, therefore, had a vested interested in open competition, the United States was unabashedly in favor of free trade during the first postwar quarter-century. Its primary protectionist device was the antidumping and countervailing duty laws. Whenever another nation attempted to sell its products in the United States at less than their market value, the threatened U.S. industry could respond by filing a petition with the U.S. Department of Commerce. That same department was then to investigate and report its findings to the U.S. International Trade Commission in order to first, determine whether such a sale had occurred and second, whether the U.S. industry had been materially injured as a result.[22] Then and only then a countervailing duty could be imposed which resulted in raising the U.S. selling price to at least the cost of production. The initiative rested with the aggrieved industry and the process was lengthy and uncertain. The steel industry did file 250 such complaints during the 1970s and 132 in 1982 alone, but found the process inadequate.[23]

Therefore, the U.S. steel industry and the USWA combined their political influence in pursuit of before- rather than after-the-fact protection. In 1968, when the U.S. Congress appeared likely to establish a quota system to limit steel imports, European producers, who were the only substantial foreign sources at the time, offered to put a voluntary limit on their exports to the United States as an alternative to mandatory limits.[24] Thus began the use of voluntary restraint agreements (VRAs). Steel imports declined modestly in subsequent years, and the VRA was

**Table 3.9**
**Consolidated Return on Steel Industry Sales: Net Income/Sales (in Percentages)**

| Country | 1975 | 1976 | 1977 | 1978 | 1979 | 1980 | 1981 | 1982 | 1983 | 1984 | 1985 |
|---|---|---|---|---|---|---|---|---|---|---|---|
| United States | 4.3 | 3.3 | 0.4 | 2.5 | 2.1 | 3.0 | 3.9 | -6.8 | -9.1 | -0.1 | -5.9 |
| Japan | 0.6 | 0.9 | 0.6 | 11.8 | 3.3 | 2.9 | 1.9 | 1.5 | -0.3 | n.a. | n.a. |
| West Germany | 1.4 | 1.1 | 0.1 | 0.3 | 0.4 | 0.6 | n.a. | n.a. | n.a. | n.a. | n.a. |
| France | -15.9 | -10.7 | -23.5 | -14.0 | -10.1 | -11.5 | n.a. | n.a. | n.a. | n.a. | n.a. |
| United Kingdom | -10.8 | -3.1 | -14.1 | -9.4 | -17.6 | -22.6 | n.a. | n.a. | -7.2 | n.a. | n.a. |
| Italy | -4.0 | -5.9 | -17.6 | -13.4 | -8.3 | -20.4 | n.a. | n.a. | n.a. | n.a. | n.a. |
| Belgium | -7.9 | -2.6 | -13.9 | -10.2 | -2.5 | -9.4 | n.a. | n.a. | n.a. | n.a. | n.a |
| Canada | 6.0 | 4.5 | 4.9 | 6.2 | 7.1 | 7.4 | n.a. | n.a. | n.a. | n.a. | n.a. |
| South Korea | n.a. | n.a. | n.a | n.a. | n.a. | n.a. | 2.6 | 4.5 | 1.5 | 1.3 | 1.2 |

*Sources: World Steel Dynamics*, "Financial Analysis of International Steelmakers"; for South Korea: Korean Iron and Steel Foundation, *Statistical Yearbook*, 1986 (Seoul, June 1986), pp. 275–302.

extended in 1972, only to be dropped in 1973 when the worldwide demand for steel boomed while the metal was in short supply. When the world steel market collapsed in 1975, sending all producers scurrying for markets and most targeting the United States, the antidumping provisions were reinitiated with the same unsatisfactory result.

Therefore, the Carter administration introduced a trigger-price mechanism (TPM), requiring government initiative primarily aimed at the Japanese threat. A set of minimum trigger prices were established based on estimates of foreign production costs, sales below which would automatically initiate U.S. government investigation and, where appropriate, initiation of expedited antidumping cases. However, the trigger prices were based on Japanese production costs. This system seemed to work satisfactorily during 1978 and the first half of 1979 when world demand was again on the rise, but when another world recession was generated later that year by the second major OPEC oil price increase, it turned out that the less efficient European producers could sell at prices above the Japanese-established TPM but still substantially below their costs of production.

The U.S. industry therefore bypassed the government TPM mechanism and returned to filing its own antidumping charges based on country-by-country and company-by-company production costs. To offset that trend, the U.S. government revised its TPM process and steel imports slowed. However, by the time a recession and reinvigorated imports arrived in 1981, a free trade administration had taken over in Washington. The TPM process was abandoned and the industry returned to its own antidumping filings. Those suits proved effective, finding European subsidies ranging from 15 to 40 percent. Direct intergovernment negotiations in 1982 substituted a voluntary agreement limiting European imports to 5.44 percent of the U.S. market through the end of 1985 in return for the U.S. industry dropping its suits. With the TPM gone and the European problem solved, the focus returned to Japan and the developing countries.

By 1984, the Japanese steel industry had captured nearly 15 percent of the U.S. market as well as other non-European nations, excluding Canada, totaling another 10.2 percent for a combined foreign steel penetration of 26 percent. In July 1984, when the U.S. International Trade Commission responded to renewed U.S. industry antidumping charges with a recommendation that the president establish steel import quotas, President Reagan refused but did agree to negotiate new bilateral voluntary restraint agreements as a substitute. The results in late 1984 and early 1985 were five-year VRA agreements with Japan, South Korea, Brazil, South Africa, Spain, Mexico, and Australia with an overall target of approximately one-fifth of U.S. steel consumption. Absence of VRA agreements with a number of producers, such as Canada, and transshipments from VRA countries to non-steel producers which resold to the United States, kept the import pace above one-fourth of the market for the first two years and above the one-fifth target until restored world prosperity reduced import pressures after 1987.

In 1988, as the expiration of the existing VRA approached, the industry began making a case for another five-year extension. Steel users opposed it, and the costs to consumers and taxpayers began to recieve some notice. The Congressional Budget Office estimated that foreign competition had cost the U.S. steel industry $2.4 billion in lost profits and the U.S. steelworkers $2.6 billion in lost wages between 1977 and 1983.[25] But at the same time, the relatively mild VRA designed to hold foreign imports under 20 percent of total U.S. consumption while actually restraining them to 26 percent was costing the nation's steel consumers over $6 billion a year in higher prices.[26] In subsidizing their own steel industries, foreign taxpayers and consumers were also subsidizing the American consumer. Since President Bush had made a campaign promise of a VRA extension, a 2.5-year extension completed the experiment.

But the aggressiveness of foreign producers in response to worldwide over-capacity was not the whole story concerning the impact of public policy. Attempts in the mid-1980s to redress the value of the dollar focused on exchange rates with Japan and West Germany, whereas South Korea and Brazil, the growing steel import threats, were relatively unaffected. At any rate, it was too late for the U.S. steel industry. Whereas in the past the industry would have responded to its lowered profit rates and the need for new investment by raising prices letting steel consumers determine how much steel they were willing to purchase at those prices, the ready availability of foreign steel denied American producers that freedom. Their pricing policies were narrowly restricted by import prices, as Figure 3.3 illustrates.

### Disinvestment

Since the challenge of low-wage, international competitors should have been obvious by the 1970s, one might have wondered

**Figure 3.3**
**Real Domestic and Import Carbon Steel Prices from 1973–1983**

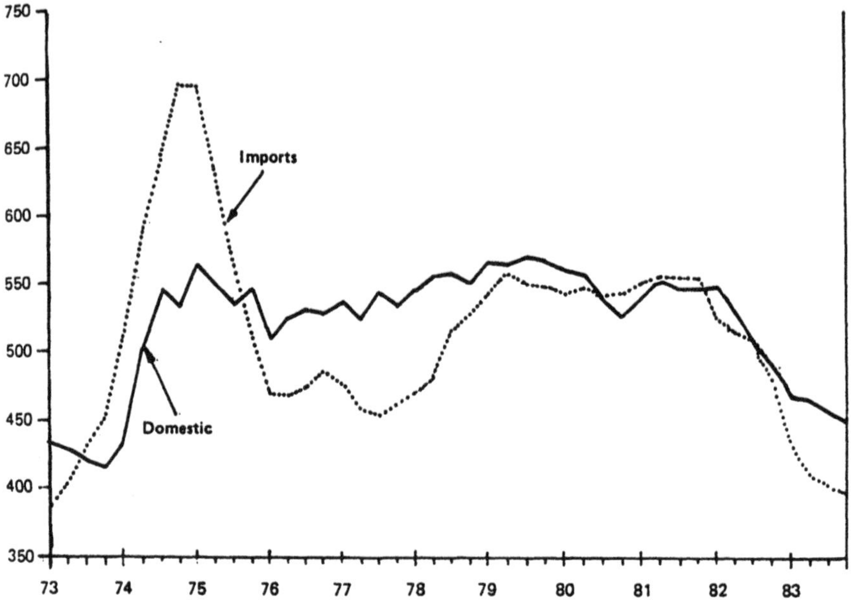


over the following decade whether steel employers were blind to the consequences or whether, by making no initial efforts at wage restraint and little toward productivity increase, they were abandoning their long-term commitment to the industry. Doing so might have been sound micro-investment policy despite the macroeconomic and social policy consequences. Steel profits were generally above the average for U.S. manufacturing during the 1950s and equaled them during the 1960s. However, they fell below them after 1974 and fell even further during the 1980s (Table 3.10). Confronted by falling profit rates in comparison to the returns on investment in other industries, steel companies began aggressively diversifying out of the steel industry, persistently pushing prices upward to generate funds for that same purpose.

But despite modest and then minimal profits, steel mills, with their high depreciation allowances and their traditional internal financing through retained earnings, were attractive "cash cows." To enhance this cash flow, the industry raised its markups in the late 1950s and 1960s while at the time it was first being

**Table 3.10**
**U.S. Comparative Return (Loss) as a Percentage of Equity for Integrated, Mini-Mill, and All Manufacturing from 1972–1983**

| | Integrated Steel | Mini-Mill | All Manufacturing |
|---|---|---|---|
| 1972-76 | 9.1 | 15.8 | 14.0 |
| 1977-81 | 5.1 | 17.0 | 16.2 |
| 1982 | (25.2) | 0.8 | 11.0 |
| 1983 | (16.5) | 3.2 | 10.6 |

*Sources:* U.S. Congress, Congressional Budget Office, *The Effects of Import Quotas on the Steel Industry* (Washington, D.C.: Congressional Budget Office, 1984), 31; Council of Economic Advisers, *Economic Report of the President,* February 1986, Table B-88.

confronted with import competition.[27] With the domestic market maturing and import competition emerging, steel stock prices fell. The combination of huge internal cash flows and the low stock prices made steel companies ripe targets for conglomerate takeovers. Five of the top twenty steelmakers were the object of takeovers in 1968 with several others threatened. Dividend payoff ratios were increased to shore up stock prices and stockholder confidence, but those were funds that could otherwise have gone into modernization to keep in step with, or ahead of the rest of the world.

While other world steelmakers were eagerly grasping the new technologies of basic oxygen, electric furnaces, and continuous casters, the U.S. industry was only moving slowly in that direction, as stated earlier. Even the new mini-mills, which used electric furnaces and continuous casters to melt scrap, which also could compete in costs with any foreign producer, and which had taken over 18 percent of U.S. steel production by 1984, were owned by maverick new entrants and not by the traditional steel companies.

At least fifteen integrated mills were declared obsolete and closed between 1965 and 1985, many of which could have been saved by modernization. A number of steel companies idled their labor-intensive steelmaking facilities, the "hot end," and began importing slabs, coils, and other semi-finished products to be processed in modernized finishing mills. Such products amounted to 5.7 percent of total imports in 1984 and even more imports later. By 1980, the American Iron and Steel Institute reported that diversification accounted for 22 percent of the industry's sales, 29 percent of all capital expenditures, and 32 percent of all assets. By 1986, the U.S. Steel Corporation, after purchasing Marathon Oil and Texas Oil and Gas, had arrived at a point where only 30 percent of its assets were in the steel industry with 26 percent of its capacity having been idled at the same time it was diversifying. Reflective of its new status as "big oil" rather then "big steel," its name was changed to USX.

However, in none of these diversifications did the steel companies prove especially successful. Steel management had nothing but money to offer its acquisitions and, in most cases, the fixed costs of the acquirer and the acquired were raised in the process. Nevertheless, the returns from investment in oil were still substantially higher than the return from the assets left in the steel industry.

Like the increased dividend payoff, diversification came at the expense of potential steel industry investment. During 1955–1980, investment expenditures of the steel industry in the United States were just equal to true replacement cost, making the net investment zero. During the early 1980s, the industry's investment fell below its replacement cost.

An industry characterized by economies of scale must maintain high capacity utilization to generate those economies. Hence, a policy of restraining the industry's size and capacity while upgrading its capabilities would have been consistent with the abandonment of obsolete facilities. A world-class capacity of around 100 million tons, instead of the 150-million-ton capacity at which the U.S. industry had settled on by 1983, after having peaked at 172 million tons in 1979, would have met the country's steel needs at competitive prices while at the same time drawing on imports only under peak conditions. At a 100-million-ton level, high capacity utilization could have been maintained with the exception of the recession years of 1982–1983 (Table 3.11).

Profits would have remained substantial. Internally generated funds could have been adequate for modernization purposes without the necessity of a mature industry competing for funds in capital markets. Despite the substantial retirement of obsolete facilities during the 1970s and 1980s, the industry became leaner but not appreciably meaner.

Looking back, had steel wages maintained their 1970 relationship to average manufacturing wages, the production cost of steel in the United States would have been $50 per ton lower

**Table 3.11**
**Output Levels, Costs, and Revenues for U.S. Integrated Facilities from 1976–1985 (U.S. dollars per ton shipped)**

| | 1976 | 1977 | 1978 | 1979 | 1980 | 1981 | 1982 | 1983 | 1984 | 1985 |
|---|---|---|---|---|---|---|---|---|---|---|
| Operating rate | 81 | 78 | 87 | 88 | 73 | 78 | 48 | 56 | 48 | 70 |
| Employment costs | 107 | 121 | 120 | 136 | 159 | 168 | 195 | 157 | 136 | 126 |
| Other costs | 206 | 224 | 238 | 263 | 302 | 328 | 386 | 373 | 376 | 359 |
| Total costs | 313 | 345 | 358 | 399 | 461 | 496 | 581 | 530 | 512 | 485 |
| Total revenue | 317 | 341 | 389 | 435 | 461 | 522 | 528 | 484 | 494 | 476 |
| Net income | 4 | -4 | 30 | 37 | 0 | 26 | -53 | -45 | -19 | -9 |

*Source:* Paine-Webber, *World Steel Dynamics*, March 11, 1986.

than it was by 1984. While average hours of labor per ton in U.S. mills had declined to 7.8 by 1982 and 6.5 by 1985, some of the best mills had gotten their labor input down to four hours per ton. Had the average equaled the best, steel costs would have been reduced another $50 per ton. These productivity increases, as noted, had come primarily from improved management, work rule changes, and employee efforts with limited contribution from technological innovation. Had the money spent in replicating traditional equipment in the 1950s and 1960s as well as the money spent on diversification in the 1970s and 1980s been invested in the most modern equipment available throughout that era, the United States could have remained the world's lowest cost *and* highest wage and profit steel producer.

Nevertheless, the Congressional Office of Technology Assessment in 1980 estimated at $5.3 billion a year for ten years the investment necessary to modernize existing U.S. capacity, while the American Iron and Steel Institute fixed at $7 billion the annual expenditure needed, including the building of new facilities. This is in contrast to the $3.4 billion which was being spent annually at the time.[28] Available internally generated funds, if invested, would have been sufficient to have replaced one-third of the existing capacity and would also have been sufficiently productive to have produced nearly one-half of the total domestic steel shipments at the time.[29] The existing steel companies would also have been the logical introducers of mini-mills for which funds were generated through external capital markets.

That is not only a conclusion from hindsight. It would have required investment of funds in the steel industry, which could readily find higher returns elsewhere (Table 3.11). The now famous comment of David Roderick, "We are not in business to make steel. We are in business to make money!" is understandable under the circumstances.[30]

## NOTES

1. Garth L. Mangum, Stephen L. Mangum, and Sae-Young Kim, "Steel on the industrial staircase: A conceptual model for early warning to other industries and nations," *Economic Development Quarterly* 2 (1; 1988): 31–49.

2. Steven Tolliday, "Competition and maturity in the British steel industry, 1870–1914," in *Changing Patterns of International Rivalry: Some Lessons from the Steel Industry*, ed. Etsuo Abe and Yoshitaka Suzuki (Tokyo: University of Tokyo Press, 1991), 70–72.

3. Rainer Fremdling, "The German iron and steel industry in the 19th century," in *Changing Patterns of International Rivalry: Some Lessons from the Steel Industry*, ed. Etsuo Abe and Yoshitaka Suzuki (Tokyo: University of Tokyo Press, 1991), 113–136.

4. William T. Hogan, *Economic History of the Iron and Steel Industry in the United States, Vol. 1* (Lexington, Mass., D. C. Heath, 1971); Mangum, Mangum, and Kim, "Steel on the industrial staircase," p. 34; Paul F. Paskoff, "The Growth of the American steel industry, 1865–1914: Technological Change, Capital Investment, and Trade Policy," in *Changing Patterns of International Rivalry: Some Lessons from the Steel Industry*, ed. Etsuo Abe and Yoshitaka Suzuki (Tokyo: University of Tokyo Press, 1991), 76–109.

5. Hogan, *Economic History*, p. 771.

6. Paul A. Tiffany, "American steel in the postwar era: Dominance and decline," in *Changing Patterns of International Rivalry: Some Lessons from the Steel Industry*, ed. Etsuo Abe and Yoshitaka Suzuki (Tokyo: University of Tokyo Press, 1991), 251.

7. Donald F. Barnett and Robert W. Crandall, "Steel: Decline and renewal," in *Industry Studies*, ed. Larry L. Duetsch (Englewood Cliffs, N.J.: Prentice Hall, 1993), 149.

8. Tetsuji Suzuki, "Import substitution and competitiveness in the prewar Japanese iron and steel industry," in *Changing Patterns of International Rivalry: Some Lessons from the Steel Industry*, ed. Etsuo Abe and Yoshitaka Suzuki (Tokyo: University of Tokyo Press, 1991), 166–190; Yoshitaka Suzuki, "The rise and decline of steel industries: A business historical introduction," in *Changing Patterns of International Rivalry: Some Lessons from the Steel Industry*, ed. Etsuo Abe and Yoshitaka Suzuki (Tokyo: University of Tokyo Press, 1991), 2, 7.

9. Seiichiro Yonekura, "The postwar Japanese iron and steel industry: Continuity and discontinuity," in *Changing Patterns of International*

*Rivalry: Some Lessons from the Steel Industry*, ed. Etsuo Abe and Yoshitaka Suzuki (Tokyo: University of Tokyo Press, 1991), 193–241; Kiyoshi Kawahito, *The Japanese Steel Industry with an Analysis of the U.S. Steel Import Problem* (New York: Praeger Publishers, 1972); Ira Magaziner and Thomas M. Hout, *Japanese Industrial Policy* (Berkeley, Calif.: Institute of International Studies, University of California, 1980), 17, 56.

10. Yonekura, "The postwar Japanese iron and steel industry," pp. 193–194.

11. Gil-Sik Jeong, "Iron and steel industry," *Korean Exchange Bank Monthly Review* (December 1983), 14–26; Y. H. Kim, *The Growth of Korea Steel Industry* (Seoul: Korea Development Institute, 1976), 21–99; Joseph J. Innace and Abby Dress, *Korea's POSCO Lights the Way: Igniting Steel* (Hunting, N.Y.: Global Village Press, 1992).

12. Marcio Jose Teixeira, *Brazil: Industrial Policies and Manufactured Exports* (Washington, D.C.: World Bank, 1983); Werner Baer, *The Brazilian Economy: Its Growth and Development* (Columbus, Ohio: Grid, 1979); Bernhard Fischer, Juan-Carlos Herken-Krauer, Matthias Luucke, and Peter Nunnenkamp, *Capital-Intensive Industries in Newly Industrializing Countries: The Case of the Brazilian Automobile and Steel Industries* (Kiel: Institut fur Weltwirschaft an der Universitat Kiel, 1988).

13. Werner Baer, *The Development of the Brazilian Steel Industry* (Nashville, Tenn.: Vanderbilt University Press, 1969), 68–79.

14. International Metalworkers' Federation, "Steel in Asia," paper delivered at the Second International Monetary Fund Asian Steel Seminar, Tokyo, in October 1985, pp. 27–29.

15. Report of the Secretary General, IISI Conference Report of Proceedings, October 1973, Johannesburg, South Africa; October 1974, Munich, Germany (Brussels: IISI, 1974) quoted in William T. Hogan, *World Steel in the 1980s: A Case of Survival* (Lexington, Mass.: Lexington Books, 1983), 15–16.

16. Robert Crandall, *The U.S. Steel Industry in Recurrent Crisis: Policy Options in a Competitive World* (Washington, D.C.: The Brookings Institution, 1981), 24; Organization for Economic Cooperation and Development, *The Steel Market in 1978 and the Outlook for 1979* (Paris: OECD), 19; *The Steel Market in 1989 and the Outlook for 1990* (Paris: OECD), 29; *The Steel Market in 1991 and the Outlook for 1992* (Paris: OECD), 27.

17. Donald F. Barnett and Louis Schorsch, *Steel: Upheaval in a Basic Industry* (Cambridge, Mass.: Ballinger, 1983), 26–29.

18. Donald F. Barnett and Robert W. Crandall, *Up from the Ashes: The Rise of the Steel Minimill in the United States* (Washington, D.C.: The Brookings Institution, 1986).

19. E. Robert Livernash, *Collective Bargaining in the Basic Steel Industry* (Washington, D.C.: U.S. Department of Labor, 1961).

20. Garth L. Mangum, Stephen L. Mangum, and Sae-Young Kim, "The high cost of peace in steel," *Challenge: The Magazine of Economic Affairs* 29(July–August 1986): 47–50.

21. U.S. Congress, Senate, Committee on Foreign Relations, *The Steel Industry, Hearings before a Subcommittee of the Senate Committee on Foreign Relations on S. 98-1002*, 98th Cong., 2nd session, 1984.

22. U.S. Congress, Congressional Budget Office, *The Effects of Import Quotas on Imported Steel*, January 7, 1985, p. 5.

23. U.S. Congress, House Committee on Ways and Means, *Problems of the U.S. Steel Industry*, Serial 98-93, 1984, p. 468; U.S. Congress, Congressional Budget Office, *The Effect of Import Quotas on the Steel Industry*, July 1984, pp. 44–47.

24. William T. Hogan, *Global Steel in the 1990s: Growth or Decline* (Lexington, Mass.: Lexington Books, 1991), 144–146; Barnett and Schorsch, *Steel: Upheaval in a Basic Industry*, pp. 239–241.

25. U.S. Congress, Congressional Budget Office, *The Effects of Import Quotas on Imported Steel*, p. 5.

26. U.S. Congress, House Committee on Ways and Means, *Problems of the U.S. Steel Industry*, p. 468; U.S. Congress, Congressional Budget Office, *The Effects of Import Quotas on the Steel Industry*, pp. 44–47.

27. Zoltan J. Acs, *The Changing Structure of the U.S. Economy: Lessons from the Steel Industry* (New York: Praeger, 1984), 137.

28. U.S. Congress, Office of Technology Assessment, *Technology and Steel Industry Competitiveness* (Washington, D.C.: Office of Technology Assessment, 1980), 309; American Iron and Steel Institute, *Steel at the Crossroads: The American Steel Industry in the 1980s* (Washington, D.C.: American Iron and Steel Institute, 1980), 43–54.

29. U.S. Congress, House Committee on Ways and Means, *Problems of the U.S. Steel Industry*, pp. 466–518.

30. Jack Metzgar, "Would wage concessions help the steel industry?" *Labor Research Review* (Winter 1983): 26.

CHAPTER

# 4 Steel Industry Joint Venture Responses

One way out of the American steel industry's financial dilemma of the 1980s was to seek foreign investment; however, foreigners were no more likely than American investors to find the steel rate of return attractive unless they had some other incentive. That incentive found its focus in joint venturing rather than either attempting to transplant outposts of foreign firms to American soil as the automobile industry had done or purchasing outright the assets of American steel producers.

## BEGINNINGS OF THE STEEL JOINT VENTURE MOVEMENT

The steel industry's joint venture movement began with raw materials and spread to steel production, which was both basic and specialty.

### Raw Material Joint Ventures

Iron ore, coking coal, and limestone, the principal raw materials for steel production, are not in short supply throughout the world. They are not, however, always available in locations that are convenient and where the quality is most desirable. After long exploitation, the traditional iron ore supplies of the European and American industries were declining in quality and ease of access during the immediate post–World War II period.

Coal and other raw materials were abundant to the American industry but not necessarily to the Europeans. The Japanese and Koreans, of course, had no raw material supplies of their own.

Canada was the most obvious alternative source of iron ore for the American industry, considering location, distance, and political compatability. However, the vastness of some of the supplies and the costliness of their development suggested joint venturing. Steel companies could engage in joint venturing with each other to spread both the costs and the risks and, thereafter, share the output. Steel companies could also share a joint venture with mining companies, with the former providing the funding and guaranteeing the demand and the latter providing the expertise.[1]

During the 1950s, American steel firms became mutually involved in fifty-three joint ventures as well as an involvement with seventeen other firms including mining, transportation, and other firms outside the industry.[2] Of the fifty-three intra-industry joint ventures, forty-four were in iron ore mining, four in coal mining, one in limestone mining, two in coke production, one shipping company, and one railroad. Bethlehem Steel Corporation, Youngstown Sheet and Tube Company, Interlake Iron Corporation, Steel Company of Canada, and Finsider and Picands Mather Company (the latter two European) engaged in an interlocking maze at different ownership rates where each controlled twenty-five of these early joint venture firms. U.S. Steel, British Steel, Armco, Republic, and other firms became involved in later years. Wabash Mines, Fire Lake, the Empire Mining Company, the Reserve Mining Company, and the Hibbing Taconite Company were examples of interindustry joint ventures extending on into the 1960s and 1970s. These steel partners locked themselves into take or pay contracts to reassure their mining partners.

Joint ventures which brought access to otherwise expensive or inaccessible raw materials were politically acceptable in the United States; that is, to everyone but domestic sources cut off from markets. Domestically, production joint ventures raised more serious questions concerning market domination while, internationally, raising questions about political consequences.

## Production Joint Ventures

International joint ventures had been attempted during the 1970s by the United States and Europeans in Australia, the Japanese and Brazilian firms in Brazil, and between Austrians and

South Africans in the latter country, but each of these attempts foundered either because of government regulation or the economic problems of the industry in the mid-1970s.[3] The political climate in the United States during those years was hostile toward any joint venture which might be percieved as reducing competition. However, the Reagan Justice Department opened a crack in the antitrust door beginning in 1984.

National Steel first raised the issue of joint ventures. While changing its name to National Intergroup in 1983, it began aggressive diversification into industries with higher rates of return, first into aluminum and then into finance. National Intergroup spun off its Weirton Steel Works into employee ownership the same year and, in 1984, attempted to sell its entire steel operations to the U.S. Steel Corporation. At that point, the Antitrust Division of the Justice Department blew the whistle, charging a serious infringement upon competition in the domestic steel industry. The sale would have been allowed, but only with the condition that U.S. Steel simultaneously divest itself of both its Fairless Works and its Granite City, Illinois plant which together amounted to about the same capacity which U.S. Steel would have gained from the National Steel merger. However, the U.S. government posed no objection when, one month later, National Intergroup announced the sale of half of its interest in its steel facilities to Nippon Kokan (NKK) of Japan. Foreign ownership of steel capacity in the United States was apparently perceived by the antitrust fighters as a step toward further diversification, the opposite direction from the increased industrial concentration which would have resulted from the U.S. Steel purchase.

Perhaps considering the implications of facilities shutdowns spreading through the steel industry but also probably reflecting the convictions of the administration, the Justice Department made a sharp right turn in its antitrust policy later in 1984. In 1977, it had allowed Jones and Laughlin Steel to acquire Youngstown Sheet and Tube, hoping to strengthen competition against industry leaders U.S. Steel and Bethlehem Steel. In 1984, after some vacillation requiring the spinoff of some facilities to maintain regional competition, the Justice Department allowed Jones and Laughlin to merge with Republic Steel. That merger involved the purchase of both by Texas non-steel conglomerate LTV, creating LTV Steel as the second largest steel producer in the United States.

Thereafter, the Antitrust Division became an enthusiastic promoter of steel joint ventures, both domestic and foreign, which were primarily committed to maintaining a viable domestic industry. Paradoxically, that promotion did not lead to further domestic mergers because all of them were too weak to be attractive partners to each other. The creation of LTV Steel, which was to find itself in Chapter 11 bankruptcy two years after its birth, ended domestic mergers in the basic industry. A decade later, Nucor, the leading mini-mill, would ally itself with U.S. Steel Group of USX by buying Brazilian ore, turning it into iron carbide in Trinidad, and then making it into steel in Arkansas by direct reduction. None of that was even on the horizon in 1984 and was still highly experimental and speculative when announced in 1994.[4] Instead, the National/NKK international partnership became the model for a rash of new alliances.

## JOINT VENTURE MOTIVES

The National/NKK joint venture reflected well the motives of their respective national industries. The American firms were the most single-minded to begin with. They wanted access to financial capital not available to them from U.S. sources. However, as they began their pursuit of partnerships, some of the U.S. firms recognized other potential gains and ended up finding even more advantages than they expected. Access to improved technologies, a greater quality consciousness, more aggressive preventive maintenance, and changed approaches to employee–employer relations emerged as by-products of the joint ventures.

From the beginning, the Japanese had essentially three motives for joint venturing with U.S. firms—they wanted access to the world's largest market. Still, opposition to steel imports were rising and the "not so voluntary" restraint agreements were a "steel fist in a velvet glove" reminder of that threat. Domestic production within the United States seemed to be the only practical opening. There might or might not be objection to "greenfield" construction of new Japanese-owned steel facilities, but, at any rate, that was not an attractive alternative. World steel capacity already had a surplus supply. In part, that was the reason for wanting strengthened and guaranteed access to the U.S. market. The only practical approach was buying into existing U.S. facilities, investing to modernize and upgrade them, and adding new facilities only where those beyond rehabilitation existed.

Finally, and most immediate, Japanese automobile company customers transplanted to the United States were complaining of available steel quality and U.S. business methods, entreating and even demanding that their more familiar suppliers come aboard the U.S. industrial scene. They had already imposed that demand upon their Japanese parts and tire suppliers which resulted in their purchase of existing facilities or the establishment of new facilities in the United States.[5] Building galvanized steel capability in the United States would have been of particular interest since the Japanese were accustomed to much heavier antitrust protection than American manufacturers because of their climate and the more intense use of ice-melting salt.

Two forthcoming international joint ventures would not fit that tri-motivation pattern. Both POSCO of South Korea and a Brazilian–Japanese joint venture named California Steel would be welcomed into that state as sources of imported semifinished steel for plants which would otherwise have closed with substantial losses of employment.

## CENSUS OF STEEL JOINT VENTURES

NKK was first with its National Steel joint venture because it had been negotiating to acquire 75 percent of the Ford Motor Company's steel division when the National Steel opportunity arose. National Steel, on the other hand, had sought to reduce its dependence on the steel industry, and a wealthy foreign partner was the next best thing to selling out. Nisshin Steel and Wheeling–Pittsburgh Steel also agreed in 1984 upon a joint venture to produce both galvanized steel and aluminum. Sumitomo Metals Industries and LTV Corporation entered into an agreement in 1986 to construct an electrolytic galvanizing line followed by a second line a few years later. Kawasaki Steel and Armco Steel entered a joint venture agreement in 1989 to operate Armco's plant in Middletown, Ohio. U.S. Steel and Kobe Steel joint ventured an operation of the U.S. Steel plant in Lorain, Ohio, in 1989, as did Inland Steel and Nippon Steel in Indiana Harbor, Indiana. By the end of the 1980s, all five of the major Japanese-integrated producers had formed joint ventures with U.S. steel firms, with Japanese firms controlling at least 25 percent of all steel capacity in this country (Table 4.1).

For the American firms, as noted earlier, access to financial support which was unavailable from domestic sources was the major attraction of these joint ventures. Investment of the magnitude required to modernize the steel industry was available in

**Table 4.1**
**Foreign Steelmakers' Investment in U.S. Joint Ventures**

| Foreign Partner | U.S. Partner | Name | Type of Operation | Location | Date Begun | Employment | Investment | Foreign Percent |
|---|---|---|---|---|---|---|---|---|
| Nippon | Inland | I/N Tek | Cold rolling | New Carlisle, IN | 1990 | 280 | $520 million | 40 |
| Nippon | Inland | I/N Kote | Galvanize | New Carlisle, IN | 1991 | 250 | $550 million | 50 |
| Nippon | Inland | | Basic | Indiana Harbor, IN | 1989 | 11,500 | $186 million | 14 |
| NKK | National Intergroup | National | Basic | Ecorse, MI; Granite, IL; Portage, IN | 1984 | 12,000 | $2.2 billion | 70 |
| Kawasaki | Armco | Armco | Basic | Middletown, OH | 1989 | 9,500 | $1.6 billion | 45 |
| Kawasaki | Armco | Armco | Galvanize | Middletown, OH | 1991 | 100 | $150 million | 50 |
| Kawasaki | CVRD (Brazil) | Calif. | Rolling | Fontana, CA | 1984 | 725 | $275 million | 50 |
| Kobe | USX Corp. | USS–Kobe | Basic & pipe | Lorain, OH | 1989 | 3,000 | $300 million | 50 |

| | | | | | | | | |
|---|---|---|---|---|---|---|---|---|
| Kobe | USX Corp. | Protec Coating | Galvan-ize | Leipsic, OH | 1992 | 100 | $200 million | 50 |
| Sumitomo | LTV Corp. | LSE I | Galvan-ize | Cleveland, OH | 1986 | 83 | $100 million | 40 |
| Sumitomo | LTV Corp. | LSE II | Galvan-ize | Columbus, OH | 1991 | 100 | $180 million | 50 |
| Nisshin | Wheeling–Pittsburgh | Wheeling–Nisshin | Basic & coating | Steubenville, OH | 1988 | 5,500 | $15 million | 10 |
| Nisshin | Wheeling–Pittsburgh | Wheeling–Nisshin | Galvan-ize | Follansbee, WV | 1988 | 100 | $96 million | 67 |
| Nisshin | Wheeling–Pittsburgh | Wheeling–Nisshin | Galvan-ize | Follansbee, WV | 1993 | 100 | $120 million | 100 |
| Yamato Kogyo | Nucor | Nucor–Yamato | Mini-mill | Blytheville, AR | 1988 | 320 | 210 | 50 |
| Kyoei/ Sumitomo | | Aurban | Mini-mill | Auburn, NY | 1975 | 315 | 300 | 100 |
| POSCO | USX | UPI | Cold roll | Pittsburg, CA | 1986 | 990 | 437 | 50 |

*Sources:* Martin Kenney and Richard Florida, *Beyond Mass Production: The Japanese System and Its Transfer to the U.S.* (New York: Oxford University Press, 1993), 157, and authors' data.

the United States only through the stock market. But the ability to float new stock offerings depended on both the immediate relative rates of return as well as the professional and public estimation of the short- and long-term outlook for the industry. The Japanese, who depended upon bank loans rather than the stock market and who also were attuned to greater long-range concerns other than fluctuating stock prices, had only to meet the banks' loan amortization requirements. The Korean government was 70 percent owner of its steel industry; therefore, govenment policies were major considerations in the investment decisions of the Korean steel industry.

With the Japanese auto transplants as a driving force, the needs of that industry were major factors in product choice. Therefore, even though it was of necessity to buy into integrated basic steel companies, the Japanese focus was primarily on a limited choice of products from those mills. Kobe Steel's choice from the Lorain, Ohio, USX mill was its high quality steel bars used in the production of engines and transmissions. Nippon Steel made a major equity investment in Inland Steel but put no money directly into that company's existing facilities. Instead, it put up 40 percent of the capital for a totally new cold-rolling mill and 50 percent of the investment necessary for two galvanizing lines on the same property in New Carlisle, Indiana. Indeed, new galvanizing capability became a major focus of the Japanese investment, as previously noted. Modernizing an existing electro-galvanizing sheet metal line near Detroit was a major project for NKK in its National Steel investment. The jointly owned National Steel subsequently became involved with Dofasco of Canada in a 400,000 ton hot-dipped line in that same country. Modernization of an older galvanizing line and the building of a new one at Middletown, Ohio, were major motives for Kawasaki Steel's investment in Armco Steel. Nisshin Steel bought 10 percent of Wheeling–Pittsburgh Steel and then put up 67 percent of the cost of a new hot-dip galvanizing line at Follansbee, West Virginia. Kobe Steel and USX invested equally in a new hot-dip galvanizing line at Leipsic, Ohio. Sumitomo Metal Industries and LTV established a joint venture in an electro-galvanizing line in Cleveland, Ohio, and another one in Columbus, Ohio.

These facilities serve U.S. as well as Japanese auto assemblers, but the Japanese were clearly the driving force. The Cleveland line of Sumitomo-LTV was designed primarily to supply General Motors and Chrysler with pure zinc-coated steel sheets. However, the Columbus line was designed to produce sheet steel coated with the zinc–nickel alloy preferred by the Japanese automobile assemblers, though it can also produce pure zinc and organically coated sheets.

Japanese auto manufacturers have not limited themselves to the joint ventures of their countrymen, even though the preference is evident as long as costs and quality are maintained. As of 1993, Honda was obtaining all of its coated steel sheets from Inland and Armco, which had joint ventures, and Bethlehem, which did not. Toyota was buying from LTV, Inland, Armco, and National Steel, all of whom were engaged in joint ventures. Nissan was buying from mills in which Nippon, Sumitomo, Kawasaki, and NKK were engaged in joint ventures. Mazda was a major customer of the LTV-Sumitomo Columbus, Ohio, facility.[6]

The auto industry was not the only motivator for international joint ventures. U.S. Steel negotiated unsuccessfully with British Steel in 1985 to import slabs as an alternative to modernizing its iron- and steel-producing facilities at its New Jersey, Fairless plant. During the 1970s, with its domestic demand booming, Kawasaki Steel had entered into a Brazilian joint venture to produce and import steel slabs into Japan to feed its finishing mills. But when steel demand slowed its growth after 1973, Kawasaki had no further need for the added slab capacity to which it had committed itself; however, Kaiser Steel had closed its Fontana, California, integrated plant. Kawasaki, Companha Vale do Rio Doce (CVRD), a Brazilian iron ore company, along with a U.S. steel warehouse owner and fabricator, the Wilkinson Group, entered into a 1984 joint venture to purchase the finishing end of the Kaiser facility for which it imports Brazilian slab. It was also in that context that U.S. Steel and POSCO of South Korea entered into a joint venture in 1985 in order to own, modernize, and operate the near-obsolete U.S. Steel finishing mill in Pittsburg, California, which was no longer able to meet the quality demands of its canning industry customers. Further, in 1989, the Japanese Yamato and American Nucor mini-mill companies joint ventured a structural mill on the Mississippi River in Arkansas capable of producing 650,000 tons a year of wide-flange beams up to twenty-four inches.[7]

## THE CONTINUING THRUST

The steel industry joint ventures were not limited to Americans involved with Japanese and Koreans. When LTV ran into financial shoals and went into Chapter 11 bankruptcy almost immediately after its purchase of the merged Republic and Jones and

Laughlin, it welcomed the spin-off of J&L Specialty Steel through a leveraged buyout in 1986. That operation supplying about 25 percent of the U.S. stainless, flat-rolled market was subsequently purchased in 1990 by Ugine, a subsidiary of France's Usinor Sacilor who was the world's largest stainless producer. A bonus from this joint venture was J&L's access to Ugine's research and development which probably included the new hot-roll-eliminating, thin-strip-casting technique being developed by Ugine with its German joint venture partner, Thyssen. Then, following a growing pattern to be discussed in the next chapter, Ugine took J&L Specialty Steel public in 1993, selling off 40 percent of its ownership in a stock market looking upon steel favorably for the first time in over a decade.[8] Meanwhile, after emerging from bankruptcy in 1993, LTV in 1995 added British Steel to its joint venture partner, Sumitomo Metals Industries, in a $450 million mini-mill to serve construction industry needs in the southeastern United States.[9]

In addition to these American joint venture involvements, Hogan lists a substantial number of international joint ventures as of 1993, along with new internal joint ventures in a number of countries not relevant to our current subject.[10] ILVA of Italy, formerly named Italsider, Kawasaki of Japan, and Siderbras of Brazil have coke ovens, an oxygen-steelmaking operation, and a slabbing mill at Tubarao, Brazil, and were adding hot-rolled strip, all for international sale of a semi-finished product. Nippon, Sumitomo, and Mitsubishi of Japan are involved with a Siamese firm named Unicord who are tin-plating and otherwise finishing Japanese semi-finished stock. The joint venture was partially funded by the Asian Development Bank and the World Bank; however, the Japanese were the prime movers. Plans were drawn up in 1991 for China Steel of Taiwan to build and take a 40 percent interest in a new integrated mill in Malaysia. While that deal failed to reach fruition, An Feng Steel of Taiwan reached in the other direction by contracting to purchase one million tons of slabs a year from a Russian plant. At the same time, the mainland Chinese government was joint-venturing with ore mining in Australia and was exploring the possibility of direct-reduced-iron (DRI) plants in both Australia and Venezuela. Japanese firms, who were constructing steelmaking facilities for the Chinese, were exploring the possibilities of joint ventures there. While Japanese, German, and Spanish interests were considering steel investments in China, a private Chinese steel firm, Shougang Corporation, was negotiating other joint ventures in the Philippines, Pakistan, Indonesia, and India. West German steel firms had already acquired or engaged in joint ventures with East German firms while Krupp, Klockner, and Thyssen of Germany were exploring a number of potential joint venture projects elsewhere in eastern Europe. Arbed of Luxemburg was involved in steelmaking facilities in Belgium and Brazil and the former East Germany while Cockerill of Belgium was joint venturing in Luxemburg and France. Italian steel firms were involved in joint ventures in Brazil, Belgium, East Germany, and Poland while, at the same time, negotiating a joint venture in China. Swedish steel firms had joint ventures in England and Finland. Usinor-Sacilor of France had joint venture interests in Spain, Belgium, and the United States.

Sumitomo Metal Industries of Japan was engaged in a 50–50 joint venture with Thai Steel Pipe Company constructing an electric pipe mill in Thailand.[11] Inland Steel had entered into a 50–50 joint venture with one of Mexico's largest steelmakers, Altos Hornos de, to sell steel products.[12] A rolling mill with a production capacity of 240,000 metric tons for bars and rods was under development in Vietnam and was owned 60 percent by Japanese interests and 40 percent by the Vietnam Steel Union.[13] The Australian government was considering construction of a DRI plant in western Australia under a joint venture with the Chinese.[14]

In addition to those joint ventures already mentioned, the Chinese steel industry was considering involvement in several other IJVs. China is engaged in modernizing much of its economy, and the demand for steel there is immense; however, its technology is behind the times. Labor is plentiful but inexperienced in modern production, and wages are low. There are also no environmental restrictions on production which would attract investment from nations where environmental concerns are major obstacles to steel production. For instance, Korea's Hanbo Steel plans to jointly build a large mini-mill in a 50–50 partnership with a local partner.[15] Toyo Steel of Japan, which has a 60 percent ownership, formed a joint venture with Shenyang Steel Rolling General of China in a new plant which would produce 240,000 tons of billets a year.[16]

The rising market for improved galvanizing for automobiles, appliances, canning, and other uses is producing a rash of joint ventures throughout the world as well as in the United States. Some of these are internal to various countries and others involve international partners. At least forty new lines are under construction in North America, western and eastern Europe, the former Soviet Union, and Japan, leading to a projected 200 lines operating throughout the world when they are completed. In addition, a number of international joint ventures are producing equipment for the steel industry. Bethlehem Steel and Usinor-Sacilor of France have a small-scale facility for the production of steel mill rolls. Voest-Alpine of Austria is involved in that country in a joint venture with Uralmasch, a Russian heavy machinery producer, developing continuous slab casters for Russian mills. The Russian company had previously constructed continuous casters in Pakistan, India, and Algeria. Usinor-Sacilor of France has purchased 50 percent of Georgetown Industries in the United States and is planning to supply technology for the production of steel rods. Germany's Thyssen is joint venturing a laser-welding operation with Worthington Steel in the United States, a thin-slab caster with Usinor and SMS of France, and another with an Italian company in Germany.

In conclusion, while IJVs involving major facilities in the United States seem to have run their course, the swelling numbers and wide varieties of smaller joint ventures throughout the world suggest that the phenomenon is not a passing one. Intensive examination of six U.S. case studies should both serve to enlighten the steel industry to its own performance as well as offer other industries a number of useful lessons from the steel experience.

## NOTES

1. William T. Hogan, *Global Steel in the 1990s: Growth or Decline* (Lexington, Mass.: Lexington Books, 1991), 205–206.
2. Daniel R. Fusfeld, "Joint subsidiaries in the iron and steel industry," *American Economic Review* 48(May 1958): 581–584.
3. Hogan, *Global Steel in the 1990s*, pp. 213–214.
4. "The odd couple of steel: Rivals Nucor and U.S. Steel may reshape how metal is made," *Business Week*, November 7, 1994, p. 106.
5. Martin Kenney and Richard Florida, *Beyond Mass Production: The Japanese System and Its Transfer to the U.S.* (New York: Oxford University Press, 1993), 155–189.
6. B. O. Huallachain, "The restructuring of the U.S. steel industry: Changes in the location of production and employment," *Environment and Planning A* 25(1993): 1347–1349.
7. Hogan, *Global Steel in the 1990s*, p. 211.
8. "Company watch: J&L Specialty Steel French connection," *Financial World*, January 31, 1995, p. 18.

9. "Company watch: LTV, think carryforwards," *Financial World*, January 31, 1995, p. 18.

10. Hogan, *Global Steel in the 1990s*, pp. 212–214; idem, *Capital Investment in Steel: A World Plan for the 1990s* (Lexington, Mass.: Lexington Books, 1992), 60–85.

11. Peter Janssen, "Thainox: Steelmakers pool their resources," *Asian Business* 29(April 1993): 15–16.

12. "Inland Steel Industries making a run for the border," *Metal Center News* 39(July 1994): 8.

13. "Hitachi Zosen: New joint venture launched in Vietnam," *Japan 21st* 38(May 1993): 10.

14. William T. Hogan, *Steel in the 21st Century: Competition Forges a New World Order* (Lexington, Mass.: Lexington Books, 1994), 39.

15. Information from Hanbo Steel Corporation in Seoul, Republic of Korea.

16. Hogan, *Steel in the 21st Century*, pp. 47–48.

CHAPTER

5

# Steel Case Studies—Three Integrated Mills

Case studies of six IJVs should provide adequate variety for many contrasting insights. Three of these involve integrated mills, and the other three involve finishing mills; one is a new facility on a new site, while the others are refurbished older facilities in long-term steelmaking locations; four are Midwestern and two Western; four serve primarily the automobile industry, while one of the others is dominated by the canning industry and the other by construction; four involve Japanese/U.S. joint ventures, one Korean/U.S., and one Japanese/Brazilian joint venture on American soil. This chapter describes the integrated mill experience, and Chapter 6 describes the finishing mills.

## THE NATIONAL/NKK JOINT VENTURE

What chance has a marriage into which the suitor entered reluctantly after denial of the the right to pursue its preferred partner, where one is constantly seeking excuse for divorce and chooses to escape at the first opportunity? However, if the result is the propagation of a reasonably healthy offspring which otherwise would have never known life, has not the union been productive?

Only one mini-mill IJV—Kyoei and Sumitomo Corporation of Japan combining with Auburn Steel in 1975—preceded the joint venture between NKK and National Intergroup of the United States in 1984. We begin, therefore, with the earliest of our case studies, based on National Steel annual reports and published sources, as well as conversations with National Steel public affairs representatives and union officials.

## NKK History

NKK began operation in 1915 producing steel tubes and pipes. Upon merger with Tsurumi Steel and Shipbuilding Company in 1940, NKK became Japan's second largest steel producer even though it was producing only 15.4 percent of Japan's total output of finished steel as compared to Japan Steel's 71.7 percent. That relationship continued after Japan's defeat during World War II, with NKK being regarded, in Japanese terms, as an obedient second son who was responsible to the leadership of the first son, thereafter designated as Nippon Steel. NKK's steel production—75 percent of its total sales—was centered at two world-class facilities; namely, the Fukuyama Works with a 16 million ton capacity and the Keihin Works with a capacity of 6 million tons. The Ogishima complex at the Keihin Works, which was completed in 1979, was considered to be one of the world's most advanced steelmaking facilities. As a leader in research and development, NKK maintained a body of 5,500 engineers to service its steel and non-steel enterprises. Though still a lagging second in size in Japan, NKK was, by the early 1980s, the third largest steelmaker in the world, employing 34,000 people with total fiscal 1983 sales of $6.1 billion.

By 1983, however, NKK had fallen victim to the worldwide overcapacity and depression in the steel business. Doubly burdened by depression in both the steel and shipbuilding industries, along with heavy interest payments on the massive debt generated to modernize Fukuyama and build Keihin, NKK lost $49.6 million. Though deeply in debt and losing money, NKK resolved to break out of this trap by entering the American steel market. No U.S. firm could have raised the funds to move in the opposite direction; however, by relying on bank loans from institutions of long-standing relationships and by being supported by Japanese government policy, NKK had no difficulty financing a joint venture.[1]

## National Steel History

The National Steel Corporation was formed in 1929 by the merger of Weirton Steel, Great Lakes Steel, and the Hanna Company. Blessed by the location of one of its two major mills on the Great Lakes adjacent to Detroit, along with other advantages, National Steel had been the only steel firm to make money during the Great Depression by operating at 61.9 percent of capacity in 1931, compared to U.S. Steel's 38 percent rate, and making $1.6 million in 1932, while U.S. Steel was losing $71 million. By being the last major steel producer to recognize the Steel Workers Organizing Committee, it maintained an intense adversarial relationship with the United Steel Workers of America thereafter.

National Steel was technologically aggressive throughout its subsequent history. By the early 1980s, it was using continuous casting for 72 percent of its steel, compared to an industry average of only 32 percent. However, the company reacted to the steel industry troubles of that period by attempting to leave the steel industry entirely. Still the seventh largest firm in the American steel industry by 1983, it had reduced its steel capacity from 12 million to 6 million tons per year while still employing 12,500 people in seven states and generating revenues of $2.3 billion from its integrated steam-shipping, iron ore, coal, and steelmaking operations. In the fall of that year, it reorganized itself as National Intergroup, Inc. (NII), and set out to diversify into pharmaceuticals, aluminum, financial services, computer services, distribution, and other business activities.

Early in 1984, NII pursued its planned withdrawal from the steel industry by selling its Wierton works to its employees under an employee stock ownership plan (ESOP) and by offering the rest of its steelmaking facilities for sale to U.S. Steel who was itself in the process of diversifying into USX. When the Justice Department refused to approve the sale for antitrust reasons, as noted in Chapter 4, NII's choices were either to abandon its steelmaking facilities and continue independent steel operations or find a buyer acceptable to antitrust enforcers. The solution was the joint venture with NKK.[2]

### Motivation for the Joint Venture

NII's motivation was simple. It had already faced the alternative of investing more heavily in the modernization of its steel facilities or of getting out of the steel business. NII had been formed to pursue the highest earnings possible and there were numerous potential investments promising higher returns than steelmaking. The Justice Department would have probably allowed a sale to one of the smaller steel firms which would not have increased industry concentration, but no viable buyers had presented themselves. Abandoning the facilities would have been a possibility if no better alternative appeared. U.S. Steel had been prepared to pay $575 million for outright purchase of National Steel's facilities and assume some of its long-term debt. NKK was willing to pay $300 million for half ownership and invest further in modernization of existing facilities as well as in the building of finishing capacity.

NKK's motives were more complex. It considered itself to be a world leader in steel technology but was unable to fully exploit its capabilities because of domestic over-capacity and rising obstacles to international trade. NKK considered the existing worldwide steel glut to be temporary and expected a growing demand in the long run. Perhaps also in NKK's collective consciousness was the fact that, while in Japan it was forced to play second fiddle to Nippon Steel's leadership, in the world outside it could exercise independence. NKK's Japanese auto and appliance manufacturing customers were just beginning to build factories and subsidiaries in the United States and were already pressing their suppliers to follow them. In explanation, NKK's President Kanao Minoru made the following statement:

> The American market is wide and deep and American politics and the U.S. economy are stable. It will take five or ten years to see whether this venture succeeds or not. But I think this venture is the first step of NKK's globalization. Through National Steel, while we can learn the American market, financial system and labour relations, we can also make friends in various industries. Moreover, we will gain the know-how to manage a business with a foreign company which is historically and culturally different from us.[3]

Patient bank-controlled Japanese capital would allow that long-term outlook, despite NKK's existing high debt load. Since NKK had spent one trillion yen to build its six million ton Ogishima complex, sixty billion yen for one-half ownership of National Steel's three million ton Detroit and St. Louis basic plants and its Chicago-area finishing mill with accompanying iron ore and coal facilities seemed a bargain. Having shared research projects for a number of years, National Steel and NKK were not strangers, making a marriage of mutual convenience an attractive step for both firms.

### The Downside and the Upside

While the joint venture appeared to its American participants as clearly the next best thing to selling out entirely, the Japanese had more formidable concerns. Yearning to leave the industry, NII had not been investing in either maintenance or improvement of its facilities. These facilities turned out to be in even worse shape than NKK had anticipated. National Steel was, of course, signatory to the United Steel Workers' contract calling for a wage level which was double the average manufacturing wage in the United States. Labor relations were as deteriorated as labor costs throughout the industry and again National Steel, with its short time horizon, had done little to improve matters. Since NKK perceived itself as competing with steelmakers in developing countries, it considered lowering labor costs to be essential. The prevailing labor relations system also appeared antithetical to the cooperative teamwork which NKK considered essential to its quality control and productivity levels. None of the Japanese automobile transplants, the previous joint ventures, or the Japanese subsidiaries in the United States had more than a few hundred employees, and most were not unionized. In other words, there were 12,000 employees organized by a strong union with an adversarial heritage.

Further, the organizational structure of the joint venture violated familiar Japanese procedures. The 50–50 split in ownership was reflected in a six-member board, three from each partner with an American chairman, a three-person Office of the President having an American president, and two Japanese vice presidents. Conflicting opinions would be difficult to resolve, not just because of cultural differences which could be bridged with the passage of time, but because one partner would have preferred, given a choice, to pull out of the industry entirely; whereas the other had made a global commitment to the future of the steel industry. Also, the location and the nationality mix among management gave the reluctant party the upper hand.

Timing also appeared threatening with steel demand still in the doldrums. However, that turned to the joint venture's advantage with the advent of new facilities as the market improved later in the 1980s.

### Joint Venture Promise

Those involved in the new joint venture did not delay in attacking these and other problems. Immediately in 1984, a nine-year, $1.8 billion capital expansion and modernization program was announced. The first phase, which concentrated on quality improvements in the primarily automotive-serving Great Lakes Division at Ecorse, Michigan, included a second continuous caster, a new electrolytic galvanizing line, a vacuum degasser to produce ultra-low carbon steel, and a joint venture with another Japanese firm, Marubeni, for ProCoil, a fabrication facility providing blanking and slitting services for automotive customers. Though of a lesser priority, improvements during the first phase were also included at the Granite City, Illinois, basic plant as well as the Chicago-area finishing mill, which was making cold-rolled and coated products.

Planned at the time and launched following the completion of this first phase at the end of the 1980s was a second phase. This phase added a second continuous caster and a pickling line at Granite City, Illinois, as well as a joint venture with a Canadian firm, Dofasco, in a hot-dip galvanizing plant in Windsor, Ontario. There were also rolling mill rebuilds, blast furnace relines, ladle metallurgy stations, and other improvements throughout the company, including a pickling line and other improvements in the Midwest Division finishing mill at Portage, Indiana. The total topped out at an expenditure of over $2 billion for both phases. Of particular interest to the Japanese partner were two new research centers in Michigan servicing primarily automotive customers. One of these centers was for technical research and the other was for product and process applications to assure a place at the forefront of new developments in quality improvement and cost savings.

Equally promising was the new Cooperative Partnership Agreement reached with the United Steel Workers in 1986 which became the pilot model of a new approach to collective bargaining and which was to spread throughout the industry during the early 1990s. It offered Japanese-style job security which meant no layoffs for those with over one year of seniority along with productivity/gain-sharing bonuses and profit-sharing in return for productivity increases and elimination of restrictive manning and work rules. The goal was a 6.4 percent reduction in labor cost to $24.21 an hour, saving $20 million annually. Since the NII alternative was probably abandonment, the USWA was amenable and, in fact, made some provisions of that agreement an objective in future relationships with other companies. However, reversing over fifty years of adversarial history was not as easy as rewriting an agreement, no matter how innovative.

The divided management was attacked in the same year (1986), with NKK buying another 20 percent of National Steel's stock from NII and Hagiwara Kokichi replacing Robert D. McBride as president. With NII in a minority position, perhaps its short-run profit orientation could be replaced by a long-term development strategy compatible with NKK's globalization ambitions.

The first five years seemed to fulfil all of NKK's expectations. Modernization took the two basic mills from 87 percent basic oxygen and 13 percent electric furnace steelmaking in 1985 to 100 percent basic oxygen by 1990 and 69 percent continuous casting in 1985 to 100 percent by 1992. Automotive customers consistently took between 26 percent and 31 percent of the firm's output along with additional amounts purchased from service centers; construction 12 to 14 percent; containers 11 to 16 percent; and pipe and tube 8 to 11 percent; with service centers declining from 28 percent in 1985 to 10 percent in 1989 before climbing back to 16 percent by 1993. However, the product line changed significantly, reflecting the rising quality requirements, particularly of the automotive and container industries. Hot-rolled sheet and strips declined from 46 percent of the output in 1985 to 34 percent by 1993. Cold-rolled sheets declined from 21 percent to 13 percent. On the other hand, more profitable galvanized and other coated products rose from 23 percent to 39 percent between 1985 and 1993, with tin mill products rising from 10 percent in 1985 to 16 percent by 1991 and tailing off to 14 percent in 1993.

In addition to pushing for greater worker involvement in decision making, the Japanese introduced strict inspection standards, preventive maintenance, and written standard operating procedures to replace the "fix it after it breaks down" and "wing it" American proclivities. Production units were regularly shut down for scheduled maintenance regardless of production pressures rather than waiting for breakdowns or malfunctions.

Maintenance and other experts were sent from NKK in Japan to inculcate the maintenance philosophy and introduce other Japanese techniques as well as substantial numbers of Americans being sent to Japan for training. Standard operating procedures were written and followed for each function and statistical process control was introduced. Whereas the best temperatures for each stage of steelmaking had never before been determined for National Steel plants, NKK experts now collected data and adjusted operations in search of optimum parameters. Operators were now informed in advance of the precise metallurgical chemistry required for various types of steel rather than indulging in trial-and-error approaches. Increased attention to quality raised the proportion of prime steel delivered at premium prices to quality-conscious customers.

On the labor front, even though no one was laid off, increased flexibility in labor deployment was indicated by the reduction in occupational classifications from eighty-six to sixteen.[4] Manhours per ton declined from 6 to 4.5. In return, in addition to job security, steelworkers had been rewarded with some $56 million in bonuses and profit-sharing by 1990; for instance, $16.9 million was awarded in profit-sharing bonuses and $3.7 million in productivity and gain-sharing bonuses, averaging $2,450 per employee in 1988; $15.5 million was awarded in profit-sharing and $9.8 million in productivity/gain-sharing bonuses, averaging $2,200 per employee for 1989.

### The Threats

All of these actions were not enough. National Steel remained a high cost operator with profit levels below the industry average. There were frequent shutdowns as well as lost time for equipment repair and modernization. A damaging fire at the company's coal mine in 1989 eventually led to permanent closing, which necessitated a substantial write-off of assets. A softening market, particularly in the prime automotive and construction sectors in the last half of that same year, required a shift of marketing efforts to lower payoff spot and export markets.

The Japanese management, in particular, had expected the Cooperative Wage Agreement, with its job security guarantees and added income opportunities, to eliminate adversarial labor relations, but they were disappointed for two reasons. First, the agreement had been reached with the two steelworkers District Directors who were deeply committed to its concepts, but that commitment never reached the shop floor where adversarial relationships remained just as intense. Relations between Americans and Japanese were excellent at the management level, but criticism of the foreign partner was also intense on the shop floor despite the absence of any direct contact. On the other hand, the steelworkers' international representatives dominated contract negotiations, and they differentiated between aggressive bargaining for contract improvements and smooth relations during contract administration. Failure to reach early agreement on a new labor contract in 1989 frightened some customers, fearful of a strike and consequent delivery interruptions, resulting in cancelation of 300,000 tons of orders.

Nevertheless, a new fifty-two-month agreement was reached with the USWA, in years thereafter bargaining hard right down to the expiration deadline. The wage concessions made in 1986 were restored, and further increases were provided for 1991 and 1992. Health and retirement benefits were improved, and the cost of living allowance was replaced, offering a unique lump-sum restoration payment of 1 percent for each percentage point of cost of living increase above 3 percent, beginning in 1991. This way, the pyramiding into the base rate experienced in the 1970s could be avoided, and the job security, productivity, and gain-sharing bonuses and profit-sharing provisions were retained. In addition, a USWA National Career Development Center was negotiated to be supported by contributions from National Steel and other steel employers, which offered internal retraining and preparation for displacement and retirement. The center was also paid for in part by overtime penalties.

The negotiation had been long and difficult but had been arrived at one year earlier than National Steel's competitors. The steelworkers' district leaders exulted the following year after 1,700 USWA members had spent 300,000 hours in advanced craft training classes by announcing:

> Benefits from the partnership approach have also accrued to USWA members working for National Steel. In past years, market softness such as National Steel encountered throughout most of 1990 would have meant layoffs or furloughs for workers. Because of the employment security offered by our labor agreement, workers with one year or more of seniority will not be laid off. No other basic steelworker in America has this protection.[5]

Nevertheless, though added income to employees had surpassed $60 million, the 1990 contribution had been almost entirely from gain-sharing bonuses of $11.2 million, and there were few profits to share. Aggressive development of teamwork with customers through a program called National Integrated Customer Service and one for suppliers called Partners in Performance had won plaudits from American customers such as Chrysler, General Motors, and Ford and Japanese transplants Toyota, Mazda, and Diamond-Star (a Chrysler–Mitsubishi joint venture). However, profits had been running below the industry average at $16.48 a ton, and the rate of return had been only 3.4 percent in 1988 and 3.2 percent in 1989, two years of relative prosperity.[6] Though sales dropped only slightly in 1990, the necessity of shifting from declining auto and construction demand to less profitable products drove profits down sharply (Table 5.1).

NKK and its supporters were becoming restive but were not ready to abandon their crusade. However, NII's desire to bail out had been reinforced, not only by the steel losses but by the fact

**Table 5.1**
**Production, Sales, Profits, and Losses, 1985–1993 (in millions of dollars)**

| | 1985 | 1986 | 1987 | 1988 | 1989 | 1990 | 1991 | 1992 | 1993 |
|---|---|---|---|---|---|---|---|---|---|
| Production | 4785 | 4955 | 5141 | 5393 | 5394 | 5735 | 5247 | 5380 | 5551 |
| Sales | 2082 | 2129 | 2417 | 2599 | 2576 | 2508 | 2330 | 2373 | 2419 |
| Net Income | -88.4 | -59.9 | 50.3 | 88.2 | 83.2 | 21.8 | -189.6 | -48.4 | -259.0 |
| Work Capital | 48.8 | 54.3 | 96.2 | 144.7 | 144.4 | 255.6 | 119.7 | na | na |
| Longterm Obligations | 386.6 | 349.4 | 320.1 | 304.9 | 280.5 | 433.5 | 485.7 | 1152.7 | 1471.4 |
| as percentage of capital | 40.1 | 40.3 | 36.0 | 31.7 | 27.7 | 36.8 | 47.6 | na | na |

*Source:* Adapted from National Steel Co., annual reports, 1989–1994.

that it lost $295 million from its total operations that year. Both companies' interests seemed to be furthered by a purchase of an additional ownership of 20 percent by NKK, providing a 70–30 split with Kokichi Hagiwara becoming chairman and Ronald H. Doerr rising from executive vice president and chief financial officer to president and CEO. Commitment to complete the capital expansion and modernization program continued unabated along with emphasis on worker training and quality consciousness. The union praised the company for keeping its commitment to no layoffs, and employee suggestions were credited with contributing to $100 million in cost savings. Partly as a result, productivity and gain-sharing bonuses amounted to $16.7 million, even without profit-sharing, raising the total since 1986 above $80 million. Employees also received their first increase in base pay in eight years. The company received $350,000 from the USWA Institute for Career Development for their enhancement courses at Great Lakes; however, they began to worry about rising health-care costs.

### The Beginning of the End

It became evident that virtue is not always rewarded. In 1991, sales were 7 percent below 1990 with a further shift to export and spot markets, resulting in an operating loss of $4 per ton shipped. That this loss, which reached $101.3 million during the first half of the year and then recovered to a profit of $22.5 million for the second half of that year, was the lowest loss per ton for any competitor was not much comfort, even though the total loss included a $110.7 million write-off of the permanently idled Mathies coal mine.

Economic sluggishness continued well into 1992. National Steel bested its competitors with an average operating profit of $25 per ton. This trend was an offset by non-operating costs such as a $50 million charge for United Mine Workers retiree medical and life insurance benefits which coincided with ceasing coal mining operations and resulted in a net loss of nearly that same amount. Savings suggested by employees helped offset the fifty cents per hour wage increase but made little dent in rising health-care costs. The addition of $14.8 million in productivity and gain-sharing bonuses, again unaccompanied by profit-sharing, raised that total above $100 million since 1986. A move from the high-cost Pittsburgh area to a new headquarters in Mishawaka, Indiana, accompanied a 40 percent salaried staff reduction which was expected to result in significant savings for the company. Nevertheless, planned capital commitments were carried out, and a new joint-ventured galvanizing plant in Jackson, Mississippi, was added.

### End of the Joint Venture

In 1993, NII bailed out and, in 1994, NKK and its Japanese financiers blew the whistle. Even Japanese capital was not that patient; besides, there were economic troubles at home, and NKK's 1992 total profit of $269 million had plunged to a pretax loss of $736 million.[7] NII had been trying to get out all along. A 1993 article in the *Far Eastern Economic Review* charged that even in 1984, while promising to maintain 50 percent ownership and plow back its profits for five years, NII had been actively seeking to dump its shares in order to bolster its drug distribution business. The same article further concluded that "no Japanese steel joint venture in the US has done as poorly as National Steel Corp."[8]

With a rising U.S. stock market, it was time to cut their losses and run; however, NKK did not abandon ship. On February 5, 1995, National Steel filed a registration statement with the Securities and Exchange Commission for an initial public offering of 10 million shares of Class B common stock. As it turned out, that was a device for NII to sell its ownership to the public, leaving NKK with a stock ownership of approximately 75 percent. As Hagiwara retired and returned to Japan, the board of directors was expanded with the new additions to the board confirming NKK's dominance.

Liabilities incurred in the closure or temporary idling of certain facilities, adoption of new accounting methods, and changes in the method of accounting for retirement obligations made it difficult to judge, from the outside, the sources or the significance of the $258.9 million loss in 1993. A new sixty-month labor agreement was negotiated with little apparent economic change. This provided for four annual $500 lump-sum payments followed by an hourly wage increase in 1995 with pension improvements and a wage reopener available at the request of either party after 1996. Further, unresolved issues were to be settled through binding arbitration. Most significant, the partnership pioneered between the union and National Steel in 1986 had now spread throughout most of the industry with added labor-management cooperation including USWA representation on the board of directors which was now added to the National Steel agreement.

As viewed from Japan, commentators were blaming American workers and American management as well as the minimal commitments of NKK's joint venture partner. American workers were still uncooperative and reluctant to work outside traditional crafts despite formal changes in the classification system. Not only had cooperation not been sold at the plant level, the USWA's international commitment to that concept did not preclude militant pursuit of economic benefits in contract negotiations. The no-layoff policy, according to these analysts, only served to embolden workers to demand "excessive" health and pension benefits.[9] As for management, Japanese analysts were critical of the retention of pre-joint venture Americans such as Doerr. Someone else suggested, "National needs its Tom Graham," referring to the U.S. Steel executive hired as Armco Steel's CEO and whom the USWA referred to as "the smiling barracuda" for his tough negotiating stance.

Then without warning on May 31, 1994, Thomas J. Usher, president of U.S. Steel, learned that the five-man leadership of its Gary, Indiana, works led by manager V. John Goodwin, had defected to National Steel.[10] (It is interesting to note that Goodwin was available because he had been passed over for the U.S. Steel presidency, whereas Tom Graham had resigned as U.S. Steel president when he had failed to rise to the top of USX. Perhaps someone at National Steel had read the *Far Eastern Economic Review* article.) Despite general recovery in the steel industry during 1993 and 1994, National Steel's losses had continued. After the number of man-hours per metric ton had declined from 5.3 in 1985 to 3.96 by 1993, the company's productivity appeared to be falling and its cost per ton shipped was still double that of U.S. Steel. Why not lure away the people most responsible for the better performance of the larger company? Doerr was out and Goodwin was in as president and chief operating officer under the watchful eye of a Japanese-dominated board of apparent impatience.

#### Summary

The joint venture was no more. The causes for disappointment were not international culture differences, but the failure of a $2 billion investment along with a ground-breaking labor agreement as well as the introduction of Japanese management methods which could have made real differences in operating costs. With the same workforce and management in the same facilities, despite modernization, the old ways seemed to have prevailed.

One year later, the scene was decidedly different; of course, 1994 had been a banner year for the entire steel industry. It was difficult to determine what to attribute to the wisdom of the new management and what to attribute to the short-term circumstances. Operations were firmly in the hands of U.S. Steel emigres. Japanese chaired and filled five of nine seats on the board of directors but provided only two vice presidents among the operating staff, one in administration and the other serving as the assistant to the president. The steelworkers chose for their board representative a lawyer who was also Allegheny County Controller rather than a union officer. More important for the short run at least, efforts were being made between management and the USWA district director to reinvigorate the desultory Cooperative Partnership, buttressing it with both short-term and long-term problem-solving teams.

The $2 billion, ten-year expansion and modernization program was complete, allowing for both reduction in capital expenditures and approval for construction of other new facilities. Pension plan credits and related legacy costs were firmed up. Further common stock sales were undertaken to reduce the 80 percent debt leverage toward a 40 percent target. At the bottom line, steel shipments rose from 1993's 5 million tons to 5.2 million tons in 1994 with sales rising to $2.7 million, which was up from the previous year's $2.4 million, the highest in a decade. The net income of $168.5 million was the first positive figure in that category since 1990.

Among the new management goals was a commitment to quality and a predictive maintenance system which seemed most closely aligned with the Japanese contributions. Most certain was the fact that, without the joint venture, there would have been no National Steel for the new management to manage. Because of that joint venture, National Steel had survived what would have been certain death.

## ARMCO/KAWASAKI

Armco Steel (originally known as American Rolling Mills Corporation), the sixth largest integrated steel producer in the United States, caught the diversification fever early even though it remained in the realm of primary metals. After decades of dalliance, it too entered an impermanent international marriage from which it emerged battered but secure.[11]

### The Perils of Diversification

In 1958, Armco purchased National Supply Company, the leading manufacturer and supplier of oil well drilling pipe and machinery. This was followed in the early 1960s by the purchase of a titanium producer, an equipment leasing company, and a producer of nonmetallic composites for the aerospace industry. However, Armco's commitment to the steel industry was evident because of its mid-1960s expenditure of $800 million to install a basic oxygen furnace and 86-inch-wide hot- and cold-strip mills at its Middletown, Ohio, facility along with electric furnaces at its Butler, Pennsylvania, works. As oil drilling boomed in the late 1970s and early 1980s, Armco planned a new $700 million seamless-pipe facility and added a continuous caster at its Ashland, Kentucky, plant to provide blooms for seamless-tube production. Then, in 1982, drilling plummeted and with it went the oil field supply business. The planned seamless mill was abandoned as was, subsequently, all of Armco's oil country tubular business, which left Ashland without an outlet for its caster production, requiring its remodeling into a slab caster. Losses in the 1982–1986 period led to the sale of some nonsteel businesses and full or partial closure of some Armco steel facilities.[12]

### The Modernization Challenge

The roller coaster ride that followed is a familiar story. After suffering through the low capacity utilization of the early 1980s to which they responded by abandoning and closing facilities, Armco found that it could not keep up with the demands of its automotive, appliance, and construction customers for flat-rolled steel as the economy expanded later in the decade. Further, its Eastern Steel Division plants in Middletown, Ohio, and Ashland, Kentucky, were no longer capable of meeting the changing specialty coating requirements of their automobile manufacturing customers whose demands absorbed nearly one-half of Armco's output.

Early in 1985, the company had launched a two-stage recovery program. The first stage had involved the sale of assets and a severe restructuring in order to cut costs along with aggressive financial management in order to reduce its excessive debt and restore lost profitability. That was to be followed by a long-term growth phase based on search for mergers, joint ventures, or acquisitions. However, unlike NII, Armco was committed to expansion within, rather than diversification away from, its familiar steel business.

The restructuring stage did allow for one new electro-galvanizing line completed at the Middletown works in 1986. Beyond that, the quality requirements of the domestic automotive companies were rising, sparked substantially by competition from foreign automobile producers, both those exporting to and those manufacturing in the United States. Foreign steel was of superior quality, and U.S. competitors were upgrading their facilities to compete.

By 1989, a five-year modernization program was blueprinted, beginning with the expenditure of $234 million at Middletown and over $100 million at Ashland. The total list of needed upgradings included

- a new continuous slab caster and modification of others
- a full continuous cold mill and needed improvements
- hot strip mill improvements
- two walking beam furnaces
- blast furnace relines and upgrades
- new coating lines and improvements
- continuous annealing line and improvements
- vacuum degassing modifications and new facilities
- shipping improvements and new facilities
- BOS bottom stirring

All of this carried a total price tag of $1.12 billion. In addition to quality improvements, the capital expenditure plans were expected to increase output by one million tons to an already more than four million annual total and at lower production costs. However, the company had no substantial treasury, its stock market prospects were at a low ebb, and U.S. banks were unwilling to risk substantial loans to the basic steel industry.

## Enter Kawasaki

At that point, the preference of the Japanese transplants to deal with familiar suppliers and obtain familiar product led to Armco's rescue. As noted, Kawasaki Steel, Japan's third largest, was involved in the United States only at California Steel with its West Coast, primarily construction industry, customer base. Armco was already serving Honda and Nissan, but it believed that Kawasaki could expand those relationships and open doors at other Japanese transplants.

Not only did the Japanese firm have access to investment capital, it was also a recognized world leader in several steelmaking technologies especially those connected with flat-rolled carbon steel and specialty coatings. Its research and development efforts employed 1,300 scientists and technicians. One of the world's low-cost steel producers, it was noted for extra deep drawing steels, automated continuous cold-rolling mills, and annealing lines. By utilizing laser welding and a looper at the entry end, Kawasaki had pioneered continuous cold-rolling mills. A "schedule-free" rolling concept had been initiated that permitted its hot strip mills to operate in close synchronization with preceding continuous casting units to comprise a virtually nonstop casting and rolling operation. Kawasaki also had continuous annealing lines operating in its mills. All of these innovations were, at this point, almost unknown in the U.S. industry.

In addition, similar to the joint venture of National Steel and NKK, the two firms had worked together through technical agreements over a number of years. Perhaps as a result, Armco's Eastern Steel Division had been experimenting for two years with concensus management which they believed would give them another area of compatability with the Japanese.

NKK with National Steel, LTV with Sumitomo, Kobe with U.S. Steel at Lorain, Ohio, Wheeling with Nisshin, and Nippon with Inland had already charted the path to automotive-serving joint ventures. An Armco/Kawasaki joint venture seemed to be a marriage made in heaven for both.

### The Joint Venture

Hence, in early 1989, a letter of intent was signed committing Kawasaki to initially pay $350 million for a 40 percent equity interest with further investments to be negotiated in the future for a possible 50–50 ownership. The joint venture, Armco Steel Company, L.P., was to be governed by a six-person management committee with equal representation from each partner. The objectives were declared to be the following:

- to respond to our customers' needs for more high quality, flat-rolled steel
- to accelerate the modernization plan for the Eastern Steel division
- to increase capacity in our most profitable product lines
- to better serve our existing customers
- to facilitate the transfer of technology from Kawasaki to Armco
- to provide better service to Japanese manufacturers located in the United States

Armco, on announcement of the joint venture, responded by stating:

We don't expect dramatic, overnight changes. Initially, in addition to our current modernization projects—including the Ashland caster conversion—we will start planning and engineering work on more of the projects on our modernization list. For the first two years, sales of the joint venture should remain about the same as this year's. Then, from 1991 through 1994, we should see a substantial increase in both sales and profitability as new and improved facilities start producing. Kawasaki will be both part owner and an active partner in the joint venture. As an owner, they will have a significant financial stake in the success of our business. As our partner, they will participate with us in the process of installing new technology and operating improvements.[13]

Whereas only 39 percent of the output had been continuously cast in 1989 and 54 percent in 1990, 100 percent was planned for 1993. Employment was expected to stay at the same level or increase, but pending retirements would offer new employment opportunities. Increased competitiveness would protect jobs while improving customer service. In other words, all would win.[14]

### A Dose of Reality

Unfortunately, the timing was not propitious. A cyclical downturn began before the ink was dry on the letter of intent. Another 290,000-ton capacity electro-galvanizing line was completed in Middletown by 1991 at a cost of $116 million, and a total of over $500 million had been spent by the end of 1993 to upgrade and modernize the facilities of the two plants. However, the added capacity was no longer needed, at least not for the short run.[15] Capacity utilization dropped to 68 percent in 1991 before recovering to 79 percent in 1992 and 91 percent on 1993. Net losses of $581.7 million for 1990, $251.1 million for 1991, $544.1 million for 1992, and $42.7 million for 1993 followed the joint venture.[16]

Under the circumstances, it was impossible to tell whether Armco was making progress on other critical fronts. Man-hours per ton of steel were still at 5.91 as late as 1992, a level respectable during the 1980s but now far above industry average. Quality complaints had not ended. Large stockpiles were still necessary to assure on-time delivery. Time had been short and the conditions chaotic, but none of the vaunted contributions of Japanese management styles seemed evident.

Meanwhile, Kawasaki had contributed an additional $208.8 million in 1991 and 1992, bringing ownership between the co-venturers to approximate equality. But by that time, Armco Steel had also accumulated $600 million of debt, mostly in the form of short-term loans from Japanese banks. However, over $1 billion in other long-term obligations had also accumulated, primarily from post-retirement health insurance and pension obligations to its 6,400 active employees along with those already retired. Amortization of the bank debt was to begin with huge payments in 1994. Long-term prospects were as good as any in the troubled industry, but the current situation was untenable.

### Under New Management

New management was brought aboard in June 1992 headed by Thomas C. Graham, former president of steel operations for USX Corporation and long-term U.S. Steel executive. Fabled for tough-mindedness, he had guided J&L Steel through radical cost reductions before taking over the presidency of U.S. Steel when it was losing $100 million a month. After overseeing the slashing of its payroll from 75,000 to 25,000, Graham had been passed over for USX chairman, perhaps for his "tell-it-as-it-is" manner. He had then taken over the small, bankrupt Washington Steel and made it sufficiently attractive to be bought out by Lukens Steel and was now ready for a new challenge.[17]

A new management team was brought aboard. A productivity-enhancement and a cost-cutting drive was undertaken which brought the man-hours per ton from 5.93 in 1992 to 4.01 during the final quarter of 1993. Product quality and reliability, including "just-in-time" delivery, was reemphasized, which lowered the customer rejection rate from 1.23 percent in 1992 to 0.61 percent in 1993, and 0.46 percent during the final quarter of the latter year. Japanese concepts of preventive maintenance were adopted. The hot-rolling mill, the cold-rolling mill, and the pickling line at the Ashland Works were idled, leaving only two coke batteries, one blast furnace, one basic oxygen facility, one continuous caster, and two coating lines to supplement the three coke batteries, one blast furnace, one basic oxygen furnace, one continuous caster, one hot-rolling mill, two electro-galvanizing lines, and three additional coating lines at the ninety-year-old Middletown Works. This did little for the joint venture's immediate financial straits. Losses fell drastically to $42.7 million for the year. The company's debt load was burdensome, and Kawasaki, like NKK, was having troubles at home, where Japanese banks were not willing to extend further credit for U.S.-based operations under the prevailing conditions. Fortunately, the United States was pulling out of its recession while the Japanese were still deeply mired. U.S. capital markets, which had been closed to steel firms a decade earlier, were now more responsive to the pared down, "lean and mean" steelmakers.

### End of the Joint Venture

Recapitalization was announced in a March 30, 1994 prospectus. AK Steel Holding Corporation replaced Armco Steel Company, L.P., with Armco retaining only 3 percent of the common stock in the new firm and Kawasaki owning 16 percent.[18] A common stock and senior notes offering was designed to bring in $838 million to reduce debt, meet underwriting expenses, and provide operating capital. By agreement, Kawasaki would provide the new firm's executive vice president, even though that commitment had not been fulfilled by the publication of the 1994 annual report, and one director; however, the joint venture was no more. The Japanese–American joint venture had saved the company and put it into transition to what turned out to be another independent public company, the survival of which is at the whim of uncertain markets even though guided by experienced management.

By the end of 1994, AK Steel was the recipient of accolades from the financial community committed to tout its stock, if not also from competing steelmakers. Both Goldman Sachs and Salomon Brothers had its stock on their "recommendations for purchase" lists.[19] Productivity improvements were cited, bringing man-hours per ton down to 3.71 for the year and 3.5 hours for the final quarter of the year. Improvements and modifications in a variety of facilities and practices were praised. Said the Goldman Sachs report:

> Prior to the new management, the company had serious problems in production efficiencies, quality, reliability and technical support. The company lost significant market share in the most quality-sensitive product lines. In the early 1980s we believe AK shipped nearly two-thirds of its mix to automotive, and about 20 percent to the appliance industries, the two most quality-sensitive markets. By 1991,the company was down to 35 percent of shipments to automotive and *no shipments* to the appliance market.[20]

By the close of 1994, the industry mix was back to 47 percent automotive with quality awards from both General Motors and Ford, an exclusive for flat-rolled steel from Saturn, and shipments to the top five appliance manufacturers. Shipments had risen 13 percent and sales had increased 26 percent over 1993, but that was not unprecedented in the buoyant steel market which had followed the 1989–1992 recession. More telling was the fact that operating income per ton shipped was up fivefold from $9 in 1993 to $46 in 1994 according to the company; this was the highest of any U.S. integrated producer and double that of U.S. Steel. Net income had rebounded from a $42.7 million loss in 1993 to a $257.6 million net income in 1994. The company announced its bank debt as totally repaid, its long-term debt outstanding reduced to $334 million, its pension fully funded, its shareholder equity up $1 million from the previous year, and its employees sharing $17 million in profit-sharing checks. Exulted Chairman and CEO Graham responded in this manner: "We continue to capitalize on cost, quality, and productivity improvements during one of the strongest steel markets in decades. Our results, by virtually every financial and operating measure, represent the best in the history of this company and its predecessors."[21]

The new management credited none of that accomplishment to pre-1992 reforms, and some with previous involvement expressed doubt about the accuracy of the company's as well as Wall Street's rosiest claims. Facilities had been idled and employment reduced substantially. When agreement had not been reached by the expiration of the steelworkers' contract at the Ashland, Kentucky, plant, the union had agreed to final and binding arbitration rather than take on its old nemesis from the U.S. Steel days. When the steelworkers filed to challenge the independent union at the Middleton, Ohio, works, management took such an aggressive anti-USWA stance that an administrative law judge from the National Labor Relations Board agreed with the union's charge of unfair labor practices. Whatever the winning formula, Goldman Sachs advised investors that "AK Steel is the most profitable domestic integrated company which should exhibit strong earnings momentum."[22]

## The Apparent Result

From the vantage point of the end of 1994, the outlook for the refurbished and restructured firm was as promising as any in the perpetually troubled industry.[23] The test for the new publicly owned steel company would be faced in the next cyclical downturn, but the signs were promising. In the 1994 annual report, only one Japanese name appeared on the masthead, a Kawasaki representative on the board of directors. All other operating officers were American. The joint venture was dead, but it left behind a lusty offspring where there would otherwise have only been an industrial grave. However, Japanese management and technological skills had not been the rejuvenating factor. Japanese capital had kept the firm alive, but it had been an unusually tough-minded American management which had brought about the renaissance.

## USS/KOBE

The joint venture between U.S. Steel and Kobe Steel is another marrying of two old-line companies in order to refurbish the aging child of one and the stepchild of the other. In that, it is akin to both the National Steel/NKK and Armco/Kawasaki partnerships, but it was more successful than the first and more lasting than the second.[24]

### Lorain Plant's History

The Lorain, Ohio, plant of U.S. Steel is more than 100 years old, having been founded in 1894 as the Johnson Steel Company and producing its first heat of steel about one year later in 1895.[25] The plant was originally relocated from Johnstown, Pennsylvania, because of the Johnstown flood of 1889. A streetcar owner named Tom Johnson decided to turn Lorain into a steel town with streetcar rail being the original product of the new plant. Becoming a subsidiary of the Federal Steel Company in 1901, which later became a part of the U.S. Steel Corporation and shifting to pipe production, the plant's name was changed in 1903 to the National Tube Company. Other products were added during the 1920s and 1930s, and by 1934, the plant employed nearly 10,000 men to produce more than 1,500,000 tons of steel ingots from five blast furnaces, twelve open-hearth furnaces, and 208 coke and by-product ovens.

In 1969, the first phase of a huge bar mill complex was opened in conjunction with the development of a basic oxygen process furnace, and pipe became the side product of the plant which was renamed the Lorain–Cuyahoga works. By the time of the 1989 joint venture, the plant had two bar mills, one of them being ten inches and the other being twelve inches, along with two seamless-pipe mills making small-diameter pipe employing about 2,600 people. The fact that 75 percent of the bar business is automotive related was the primary cause of the joint venture.

### Kobe Steel's History

The tiny Kobayashi Steel Works started steel casting in 1905 in the small village of Kobe with one 3.5 ton open-hearth furnace imported from Great Britain.[26] The first owner, finding himself unable to obtain the huge capital investment needed for expansion, sold the company to another who changed the name to Kobe Steel Works and became, in 1909, the designated steel supplier to the Japanese navy. Kobe Steel thereafter specialized in producing ship parts such as anchors, crankshafts, and propeller shafts. They continued to expand in order to meet the World War I demands for steel forging and casting and thereafter diversified into general steel and machinery, but with a continued strong connection to the Navy.

By 1935, Kobe Steel had become the third largest steel producer, supplying 7.4 percent of the total output of Japan and 42.7 percent of the nation's wire production. During the Japanese First Rationalization Program (1951–1953), Kobe Steel invested 1,400 million yen to expand its steelmaking capacity with a basic oxygen furnace and a continuous caster. Following the two oil crises, however, the Japanese iron and steel industry was unable to recover to the growth rates of the 1960s. The industry's output and profit stagnated during the 1970s and 1980s, and the competition with steel from developing countries and mini-mills had become fierce.

In 1983, all Japanese integrated steel companies lost between $40 and $57 million; and all except Kobe Steel suffered a decrease in sales. Kobe accomplished this not from steel production, despite its 20,000 employees engaged in that endeavor, but from increased engineering sales and equipment exports. Therefore, it was ripe for any opportunity to break out of Japanese market limitations and, in fact, had established facilities in Singapore, Venezuela, and at locations in Africa. Some of these ventures were wholly owned and some were joint ventures; some were basic mills and others were mini-mills.

### Motivation for the Joint Venture

Committed to diversifying away from the steel industry and on its way to becoming USX, the Lorain plant's parent company refused to put any money into it after 1980. Overall, the mill was profitable, but its entire profit came from its high quality tubular products. Whereas U.S. Steel once had eleven bar mills, Lorain's was the last one and the corporation was anxious to divest itself of it. Unless a buyer could be found, the Lorain works would be allowed to die a slow death. U.S. Steel went looking for a purchaser and received many prospects from 1987–1988 before it struck its joint venture deal with Kobe Steel in 1989.

As a major supplier to the Japanese automobile industry, Kobe Steel was under pressure from its customers with transplants in the United States to relocate there as NKK, Kawasaki, Nippon, Sumitomo, and others had already done. Kobe Steel was not without reluctance because, to the extent they produced within the United States instead of exporting to it, Kobe Steel's domestic output would suffer. But during the formative period of its decision making, the voluntary restraint agreements were in force, and it was uncertain whether imports into the United States might be cut off entirely. Therefore, Kobe Steel had an interest in getting into the U.S. steel market. Besides, it was considered a learning opportunity. Kobe Steel was an expert in wire and bar and experienced in most flat-rolled products, but it did not make any seamless steel and had no tubular experience. At the Lorain plant, it could make steel bar for the engines and transmissions of the Japanese auto firms while, at the same time, learning tubular techniques and markets without a big investment.

As the deal was struck in 1989, Kobe Steel paid $300 million for a 50 percent interest in the plant. U.S. Steel provided the plant for its half interest, placing the $300 million into its general fund rather than putting it back into plant improvements. The parties agreed that $500 million was needed for refurbishing and modernization, $100 million being needed immediately for a blast furnace to be rebuilt. Kobe Steel was willing to put up its share of the additional money but USX was not. Any additional funds would have to come from the cash flow of the plant itself. The parent companies did agree to leave all of the revenues within the joint venture for the first five years and only take enough out in dividends in order to pay taxes.

### Modernization

Projects which were on the books at the time that the joint venture was formed included the upgrade of two blast furnaces and the rolling mill and installation of a continuous bloom caster, the latter having a capacity of 100,000 tons per month and expected to cost $64 million. Along with the existing billet caster, this would double the plant's continuous casting capability to more than 90 percent of its output. The joint venturers also committed themselves to increase competitiveness by modernizing the facilities in order to increase production efficiency and produce higher value-added products. Added to these plans were modernization of the coil processing facilities and plans to build a seamless mandrel mill. Though these new and upgraded facilities were to be paid for from plant revenues, the joint venturers did agree to build a 100,000-ton-per-month hot-dipped galvanizing facility at a cost of $200 million which was to be shared equally.[27] The plant had established its own long-term financing as well as lines of credit to meet payroll and other current operating costs; however, capital investment must be in cash, not credit.

The major products of the basic plant continue to be bar and rod, primarily for automobiles, and tubular products for the oil and gas industries.[28] The new galvanizing mill would also primarily serve the automobile industry.

### Technological Progress and Productivity

The Japanese reputation had raised the expectations of the Americans so high that it was somewhat of a disappointment to learn that they were just regular people after all. Since the Japanese had been considered the world leaders in the bar market, a remarkable transfer of technology was expected; but the Japanese turned out to be only modestly ahead of the Americans in that technology.

After some experience, U.S. Steel veterans at Lorain concluded that quality was a religion to the Japanese but only a practical goal for themselves. They had been already aware of the advantages of Japanese technology, but it had been out of their reach financially. Quality consciousness was more of an attitude rather than a technology or financial investment and could have been within American reach from the beginning. A major contribution of the Japanese was to help the Americans focus on that objective. A commitment to total quality management (TQM) has been made with everyone in the plant supposedly dedicated to do their job to the highest quality standards. The Japanese attention to detail is described as phenomenal, including a process discipline that Americans admit they lack. The Japanese rarely have the problem of people failing to follow procedure. While Americans worship heroes and promote individualism, the Japanese operate as groups; they follow procedures while Americans take shortcuts. Americans are improving in these areas but are still behind the Japanese.

This does not mean that the American management at Lorain was unaware of the need for change. The zero defects program and other quality efforts had been introduced within U.S. Steel during the 1960s and 1970s. However, they proved to be false starts either by being given only lip service or by being abandoned at the peaks of demand when everything made could be sold or at the troughs when money was scarce. Consequently, the plant was remanned after the 1986 shutdown. Many of its jobs were grouped together, eliminating narrow classifications and allowing for much greater flexibility in training people and moving them around in broad categories. Whereas there had been twenty-seven different job classifications in the open hearths, now there were only four. As the plant opened again after the shutdown, the union agreed to all of that flexibility because everyone was afraid of a permanent closure. But whether that flexibility would have survived the next bout of either prosperity or recession is not known.

Employee participation in a team environment had also been launched at Lorain before the Japanese came. Rick Mahon, human resource manager at the plant, says he would not have come to Lorain from outside of the steel industry had he not been guaranteed the right to introduce team participation. However, he admits that the team approach, and therefore his tenure, might not have survived over the long run had it not been for the arrival of the joint venture not long after.

Training is another example. In the past, as soon as there were losses, the first expenditures cut were those for training. This trend came and went among other fads; now it appears to be a basic commitment and a way of life. At least, there have been no noticeable decreases in training in downturns since 1989 and with the Japanese involvement such cuts appear highly unlikely. The American management changed as a matter of necessity; otherwise, the plant would have closed. Under the joint venture, these are not programs, they are basic philosophies. However, no one is relying on change of philosophy alone; for instance, quality requirements have been so entwined in the incentive program that peer pressure to maintain it is intense.

The commitments to employee participation and training and quality consciousness at Lorain have been primarily from the ground up or, in other words, from newcomers among the American plant management team and from their nineteen Japanese counterparts rather than from the top down. They have had the full support of both joint venture parents at least as long as they did not ask for money which might have been forthcoming from Kobe Steel but not from USX. The Lorain plant has been able to obtain information and consultancy from its parents, which, as a practical matter, has had to come from the more technically developed Kobe Steel. However, neither parent has put any money in the plant for new technology; the plant has had to generate its own investment.

As restraining as that may be, the results are apparent. Whereas a decade ago, the labor requirement of the plant was 6.6 hours per metric ton, for 1994 it was 3.92. However, those gains are typical across the corporation and throughout the industry where four hours per ton is the norm except for the minimills where it is half that. There is no way of proving that the greater productivity comes from participative management; but the Lorain management is convinced it does. Customer rejection data may be an even more objective measure of progress. Before the joint venture, it had an index of 1; now it is 0.2 or 0.3.

### Profits and Markets

The Lorain plant has always been in the black, even in the years that the rest of the U.S. steel industry was losing money, and it also made money in each of the 1989–1994 years following the joint venture. The tubular market in 1990 was outstanding; consequently, the plant made $7 million which allowed it to reline the blast furnace without going into significant debt. With continuing profits, the plant has been able to continue its modernization program, also without adding to long-term debt.

The plant purchases raw materials through USX from its iron ore mines in Minnesota and its Appalachian coal mines and exports nothing directly except for small amounts to Mexico and Canada. However, USX buys and resells, through export, a substantial part of Lorain's tubular production. As noted, the Japanese auto companies encouraged the joint venture and have also given their own business to the plant. Just as Kobe Steel cannot talk for the Lorain plant in Japan, Toyota cannot dictate to their plants in America. Nevertheless, the Japanese companies feel comfortable dealing with the plant and purchased more than 40,000 tons of steel from the Lorain plant in 1993. Because of that comfort level, Honda, Mazda, and Nissan, among others, are more likely to buy from the joint-ventured plant than from its competitors. For example, Copperweld was a major supplier to Honda until the withdrawal of Copperweld's Japanese partner, after which Honda, beginning to feel uncomfortable, approached the Lorain plant as a replacement. Nissan is also coming around.

However, automobile companies—foreign or American—do not necessarily buy from the Lorain plant directly. If a crankshaft is to be made from Lorain steel, the steel might be sold to the parts manufacturer which would then sell it to the auto company. Nevertheless, Japanese parts manufacturers in the United States are influenced by their Japanese customers. Besides, the final user often specifies where the raw materials in the parts are to come from. In fact, the trend is for single billings with the steel plant coordinating traffic to the final customer. Either the crankshaft is ordered from the steel plant which contracts with a forger for turning the steel into crankshafts, or the automobile manufacturer orders crankshafts from the supplier which, in turn, buys the steel.

### Cultural Differences

Cultural differences have been a minor problem between the U.S. Steel and Kobe Steel representatives at the plant. Of course, English language competence varies among the Japanese, but they struggle to learn. There are also the usual differences in attitudes. The Japanese are generally not as verbal as the Americans and tend to suspect an inverse relationship between a lot of talk and intelligence. Like other Japanese in the steel industry, those from Kobe Steel are addicted to long-term planning and have a much longer time horizon than their counterparts. The Kobe Steel representatives are committed not only to preventive maintenance but also to predictive maintenance—changing parts at regular intervals rather than to wait and monitor them until the part has become noticeably worn. The Japanese do not wait until a bearing makes a noise; they change it on schedule, regardless of its continued usefulness. Cleanliness is also more important to the Japanese. Susumi Okushima, the Japanese executive vice president, upon his arrival at the Lorain plant, reported to his superiors at Kobe Steel that the Americans at the plant were living in dirt. It was not the quality of the output—he considered that very good—it was the housekeeping that jarred him.

However, the learning has not been one-sided. The designated term of service for the Japanese at the Lorain plant is five to seven years. Ten of the original complement of nineteen were still in the United States in 1994 but were expecting to return to Japan in the near future. Upon reflection, they agreed it had been a learning experience for a career in international business. These same Kobe Steel representatives believe they will take much new knowledge back with them. For instance, Japanese steelmakers tend to think that once they start a steelmaking process they can never stop it. In the United States, because of strike-prone labor relations and business cycles, steelmakers are accustomed to banking furnaces and idling while still maintaining facilities. That may become important knowledge for the Japanese; even though they do not expect to become strike prone, their current recessions suggest they may have to accommodate to the business cycle. They hope to have learned how to make rational investment decisions, something they say the Japanese have not been good at, while also learning how to minimize investment and maximize return. They further reflect that Kobe Steel has tended to hire too many people and expect to operate in a lean fashion when they are in a position to influence manning.

### Participative Management

The participative management system was already in place at the plant before the joint venture but was extended and improved thereafter. An American vice president believes there would never have been as much participation had not the joint venture followed shortly thereafter. Since the joint venture with Kobe Steel, the planning horizon has changed dramatically. Where the plant used to have an annual plan and was not generally taken very seriously, it now has a formal, very detailed five-year plan with a tracking and analysis capability. The Japanese taught the Americans how to set up goals and accomplish them in participation as well as in production. It was the Japanese who prepared performance indicators as well as plans and goals at the design and measurement levels rather than the implementation level. They did not hold formal training sessions for the Americans but taught them how, one on one. By undergirding and supporting each division manager as his advisory staff and by participating in the various teams, the nineteen Japanese proved to be a potent force for change.

Team leaders may be hourly personnel or former supervisors, but they are all working members of the crew. A seventeen-year supervisor in the blast furnace who admits to having been a cynic when the team leader program was presented, now confesses to being a convert. Peer pressure keeps people producing. There are resource people available for technical advice. The team leader is like a supervisor except that he is also a working member of the team. They are chosen by management, at present, with no union involvement in the selection process.

Management invited the union to join them in this process, but they declined.

There are still union grievancemen at the plant, but they are not to be confused with team leaders. In effect, the grievance committee member bypasses the first level—the team leader—and brings grievances to the next level of management. Contracting out is the foremost source of grievances, a subject which is not within team control.

### The Training Program

According to Robert Sabota, Manager of Training and Development, the employee participation has not been a dramatic step but an incremental one which is closely related to the problem-solving team activities. Everyone, both management and labor, has had to learn to improve practices as production continues. This has been the responsibility of an intensified training effort.

The orientation training program of the plant was originally very intensive, all on company time and usually on overtime, after or before the regular shift. Late in 1994, the plant was doing refresher training taking place on one-half days, once every other month. Employees, managers, and suppliers had been combined in groups of about forty-five with about 5 percent of the workforce in training at any given time. Quality consciousness, recognition of threats to quality, and problem-solving abilities were emphasized. Social concerns are not neglected, however. These range from customer relations concerns to awareness training on issues such as sex and race harrassment internally, as well as external environmental concerns. Most of the training programs were developed before the joint venture, but the Japanese have reinforced them and, it is supposed, guaranteed their future.

The plant has engaged in considerable cultural training for both national groups. With the turnover, it is hard to establish a long-term relationship; it is, however, mostly an effort to help the Japanese cross the cultural bridge. Communication skills are not highly valued by the Japanese. Their typical stay is one to three years, and they do not necessarily come by choice. Therefore, it is more practical to train the Americans to understand the Japanese.

### Personnel Practices

Compared with other joint American and Japanese ventures such as National Steel or Armco, the human relations side of the

Lorain plant joint venture seems to be at least one of the keys to its success. Money can take care of the technological matters, but in order to establish sound cultural and human relations much more is required.

The Japanese complement consists of the executive vice president, one other manager, and seventen technical people. Only the executive vice president had been at the Lorain plant for the full 1989–1994 period with the others having experienced considerable turnover. Basically, the Japanese at the plant were there on an advisory level only and were not involved in direct supervision. The Japanese, who are selected and supervised by the Japanese executive vice president, are there as technical support dotted lined to the division managers. However, an effort is being made to get the Japanese more involved, especially in quality control.

Nevertheless, the influence of the Japanese permeates far beyond their number. The seventeen technical people continue to work closely with and for the division managers. They are not isolated and working together as a group but are mixed among and working directly with the Americans. Through daily discussion and routine conversation, their assignment has been not only to teach their engineering techniques but their problem-solving methods; namely, how to accomplish goals; set targets; chart progress; and, if necessary, change methods and practices. Plan, execute, check, revise if necessary—these are all part of the classical Deming circle exported to and reimported from Japan. As a result, the influence of the nineteen Japanese seems to permeate the whole plant of 1,700 workers to a surprising degree. Every day, each of those Japanese people have conversations with key managers as well as lower level employees. Through everyday conversation, they spread their influence. There has been no attempt to measure this influence, but it is probable that no one in the plant is unaware of it, or unaffected to some degree. The Kobe Steel complement did not, but probably should have, include human relations managers, but, with the receptive environment from the American human resources staff, the impact has been much the same.

The Japanese are not involved in personnel selection. Selection criteria have been changed to look for team and participation orientation, among other criteria, in new hiring situations. The plant has developed some psychological tests to look for participativeness and, judging from the probationary period performance of twenty-six new employees, the human resources staff is convinced that the tests were effective.

Rick Mahon, General Manager of Human Resources at Lorain, reflects:

I can't measure what our new selection process has accomplished. I can tell you one thing though. I wouldn't have come here in 1990 had there not been a joint venture, because I had a previous taste of what the steel industry was like. Basic steel had a reputation as being confrontational, not the most progressive industry in the world. I at least had a perception of what the Japanese influence on a company was in terms of their environment. I felt strongly from talking with my predeccesor, who used to be the general manager of human resources, when he hired me, that this was a different environment than the traditional steel industry environment. I had worked a lot in those metal industries and appliance plants where participation was underway. Everywhere I went there was real importance placed on the relationship between the company and the union and between the employers and the workers. There were a lot of participative things we did to make that happen. I really did not want to go back into the adversarial arena. The fact that this was not a typical basic steel plant and that the Japanese were involved made it attractive to me. I would not have come otherwise.[29]

### Pay Practices

The Japanese feel strongly about variable compensation tied to performance. With Japanese encouragement, the Americans are trying to expand the variable compensation component of pay and design and implement measures which relate to the accomplishment of long-term objectives as opposed to short-term business plans. They are leaning toward profit-sharing, accompanied by incentives determined by the accomplishment of measured and quantified objectives. Fortunately, the union's thinking is similar, so no clashes are anticipated. A five-year plan has been introduced in an attempt to get away from immediacy and reach other longer run objectives with attention to such subjects as safety and targets on return on assets. The plant was about to introduce a gain-sharing plan in order to motivate people to change the way they work. Accompanying the gain-sharing plan was a plan to reduce the workforce through normal attrition with hopes of getting down to three man-hours per ton. This is doable only if the reduced numbers are enough to get the job done without overtime. Accomplishing this is a slow process.

Gain-sharing plans could be introduced without negotiation if they did not conflict with the union agreement and their plans did not. The Lorain plant has seventy or more incentive plans into which it negotiated the right, in 1991, to introduce quality variables. That has reaped some significant benefits for both management and employees. Two of the plans have returned three to five times the amount of incentive increases previously obtained by the employees as a result of these quality elements. Management estimates that it is about halfway to where it can legitimately claim quality variables as its basic approach to improving production per man-hour. The gain-sharing and quality-based incentive process is not attributable to the Japanese influence. This is probably what the Americans would have done anyway, but it is certainly consistent with the Japanese thinking and supported by them.

### Labor–Management Relations

The local union at Lorain had traditionally exercised little influence. Contracts were negotiated at the international and corporate level, and the local union played a small role except for the early stages of the grievance procedure. However, this had begun to change just before the joint venture occurred with the introduction of employee participation, a development that the Japanese were happy to find in place without having to take responsibility for it.

The Japanese influence has been much less evident with labor relations than with personnel practices. Kobe Steel had one strike right after the war, but both the company and the union were hurt; consequently, both decided against further militancy and never had another strike. The company has made decisions without discussion with the union to an extent that would be unacceptable in an American mill. Lifelong employment has been a factor in complacency; however, this is changing. Now the company often asks someone to leave, which makes the system more confrontational. The Japanese try to avoid confrontation and achieve concensus, which often takes a long time and a lot of energy. Further, there are no seniority systems or grievance procedures in Kobe plants; increasingly, the practice is juniority. If the employer decides to ask an employee to leave, the latter has no choice. Job security is at company discretion. The Japanese admit it is more difficult to manage in an American context, but they have little to contribute on this front.

While the Japanese have stayed involved in the day-to-day practices of participation, they have concentrated on technical issues and avoided both salary and labor relations issues. They do not understand the U.S. labor relations system and, by choice, did not sit at the bargaining table in 1991 and 1994. They are not confrontation oriented and are uncomfortable with the American proclivity to draw a line in the sand; conversely, the Japanese generally pursue compromise and concensus. They support their American partners in the labor relations process but do not do so with enthusiasm.

The Lorain plant's labor relations staff has been disappointed with their labor relations experience under the joint venture despite the fact that it does not appear to be negative when viewed from the outside. Part of their disappointment has emerged from their expectations. With the collapse of industrywide bargaining after 1983, Lorain negotiators hoped to get a plant agreement tailored to what they considered to be their needs. This proved impossible in 1986 when they were still part of corporation-wide negotiations and agreement. With the joint venture, they believed a separate agreement stressing the peculiarities of their bar and tubular product lines and operating circumstances should be possible; they were again disappointed. On the labor side, the negotiations were ostensibly done by the district director who favored product line agreements. However, the negotiations occurred at a separate facility in Pittsburgh under the watchful eyes of the international union and corporate management with Bernard Kleiman, General Counsel of the USWA, as chief union negotiator as well as corporate labor relations personnel dominating from the management side. USWA President Lynn Williams came to the Lorain plant to address the workforce while the negotiations were under way; this was, in local management's view, simply window dressing. Williams's commitment to participative management and union representation on corporate boards was not evident until the following negotiation. Local management also overestimated how much of a separate company they were going to be allowed to become. They also underestimated what it would take to be able to negotiate a separate product line agreement. Negotiation responsibility would have to devolve to the plant level for that.

The 1991 negotiations were also marred, in the judgment of Lorain's human resource managers, by excessive involvement in the collective bargaining process by local champions of the participative environment. As the first labor agreement negotiated after the joint venture, there were too many unresolved issues and relationships for amicable participation. Consequently, there was serious conflict on the shop floor about what the new labor agreement was going to entail and how it should relate to the rules governing employee participation.

The major issues in the 1994 negotiations were what the union called their "twelve points of light," which were subsequently chronicled as the basis for "new directions" in steel industry collective bargaining. Counterpoised with management's desire for a long-term agreement and rather than having to face the potential of a work stoppage again in three years, the new directions were apparently accepted without wholehearted endorsement. Along with improvements in wages, pensions, and benefits, the twelve points of light included, among other things, a twenty-three-page "participative partnership" agreement mandating employee and union participation up to, and including, a seat on the management committee or board of directors.

This joint decision-making document was designed to involve the union more in both corporate policy as well as day-to-day decisions. The unanswered question during 1994 site visits was how that newly agreed-upon formal framework for participation would impact upon, overlay, or replace the extant, unilaterally introduced system which functioned with the approval of, but without the direct involvement of, the local union hierarchy. Employee representatives had been separately chosen for participation which continued separate from the union grievance procedure and other aspects of contract negotiation. That unilateral approach would have to change, but the nature of the change was not yet apparent.

The Lorain plant management was pleased with the sixty-six-month length of the agreement reached in 1994. They complained that union negotiators were still committed to negotiating an umbrella flat roll-dominated agreement with U.S. Steel and then, as local management saw it, "jamming it down everyone else's throat" rather than negotiating a separate Lorain agreement. The international union's negotiators, however, saw the demand for a product line agreement as a request for concessions. The union's commitment was to establish a "level playing field," accepting variances from the industrywide pattern they were pursuing and the companywide agreements they were insisting on only for firms which were losing money. The Lorain plant was not in that condition and, therefore, should not be treated differently than other basic steel plants. With greater responsiveness from corporate negotiators, there were some elements in the 1994 agreement that were different from the U.S. Steel agreement, recognizing the uniqueness of the operation, but not to the degree that local management would characterize it as a product line agreement.

### Summary

The Kobe–U.S. Steel joint venture at Lorain, Ohio, has been a highly successful one. The sources of that success have been an open market for bar and tubular steel which was not flooded by competitors; an added facility for demanded galvanize, a strongly established position in the automotive industry augmented by favorable treatment from Japanese auto transplants; a Japanese purchase not characterized by heavy fixed debt so that profits could be set aside for modernization; an American co-owner prepared to stay the course; a plant staff already moving in the directions that their new Japanese partners would emphasize; and a union who, if not positively involved in the new policies, were at least not in opposition. With this happy combination of circumstances, the relationship should be long and profitable.

## NOTES

1. Seiichiro Yonekura, *The Japanese Iron and Steel Industry, 1850–1990* (New York: St. Martin's Press, 1994), 263–272.
2. William T. Hogan, *Global Steel in the 1990s: Growth or Decline* (Lexington, Mass.: Lexington Books, 1991), 19–21.
3. Quoted in Yonekura, *The Japanese Iron and Steel Industry*, p. 268.
4. "National Steel puts its 'Detroit' strategy in gear," *Iron Age* 4(July 1988): 25–29.
5. As reported in the 1991 annual report.
6. George W. Hess, "Look behind Doerr no. 1 at National Steel," *Iron Age* 6(November 1990): 32–33.
7. "Down and dirty in the steel belt," *Business Week*, June 20, 1994, p. 44.
8. Susumu Awanohara, "Nerves of steel," *Far Eastern Economic Review*, November 11, 1993, p. 57.
9. Ibid.
10. "Down and dirty in the steel belt," p. 44.
11. The Armco/Kawasaki story was reconstructed from information provided by company public relations staff, conversations with Kawasaki personnel, discussions with USWA representatives, and public sources.
12. Hogan, *Global Steel in the 1990s*, pp. 17–18.
13. Undated news release, *Armco & Kawasaki Steel Announce Joint Venture to Improve Specialty Carbon Steel Operations*.

14. Ibid.
15. Hogan, *Global Steel in the 1990s*, pp. 208–209.
16. CS First Boston, *Prospectus: 17,572,222 Shares AK Steel Holding Corporation Common Stock*, March 30, 1994, pp. 32, 33, F-6, F-7.
17. Erle Norton, "Chairman of AK Steel tries to shake off tag of 'operating man,'" *Wall Street Journal*, November 25, 1994, p. 1.
18. CS First Boston, *Prospectus*, pp. 14, 36.
19. Anthony H. Carpet, *AK Steel Holding Corporation*, U.S. Research, Goldman, Sachs & Co., New York, October 18, 1994; Michelle Galanter Applebaum, *AK Steel—Living Up to Its Promise*, United States Equity Research, Salomon Brothers, New York, December 5, 1994.
20. Carpet, *AK Steel Holding Corporation*, p. 10, and 1994 Annual Report.
21. Remarks of Thomas C. Graham, Chairman and Chief Executive Officer, AK Steel to the NAPM-Steel Buyers Group, Metropolitan Club, Covington, Kentucky, November 8, 1994.
22. Carpet, *AK Steel Holding Corporation*, p. 10.
23. Remarks of Thomas C. Graham, Chairman and Chief Executive Officer, AK Steel to the NAPM-Steel Buyers Group, Metropolitan Club, Covington, Kentucky, November 8, 1994.
24. This section is based on site visits to the Lorain, Ohio, plant and interviews with both American and Japanese management as well as local and international union representatives along with published sources.
25. U.S. Steel Corporation, *Steel Town: A History of Lorain and Lorain Works* (Lorain, Ohio: Lorain Works—USS, 1980).
26. Yonekura, *The Japanese Iron and Steel Industry*, pp. 62–63, 185–186.
27. "Kobe Steel–USX joint venture to install bloom caster: Investment to lead to improved production efficiency," *PR Newswire*, June 7, 1994.
28. B. O. Huallachain, "The restructuring of the U.S. steel industry: Changes in location of production and employment," *Environment and Planning A* 25(1993): 1348.
29. Interview of Rick Mahon and staff by Garth Mangum and Sae-Young Kim, August 1, 1994, and subsequent discussions with USWA representatives.

CHAPTER 6

# Steel Case Studies—Three Finishing Mills

The finishing mill experience is in sharp contrast to that of the basic mill joint ventures for reasons that will become apparent. California Steel, UPI (for USS/POSCO) and the combination of I/N Tek and I/N Kote are our examples.

## CALIFORNIA STEEL INDUSTRIES

California Steel (CSI) is unique in that it is a joint venture of Japanese steelmaking and Brazilian iron ore interests on American soil without American ownership.[1] A metaphor for this relationship could be an expatriate couple in U.S. residence were it not for the fact that the joint venture involves only a handful of non-U.S. managers and employees.

### The Origin of Kaiser Steel

The only integrated basic steel mill ever on the West Coast of the United States, Kaiser Steel was created at Fontana, California, early in World War II by contractor-turned-industrialist Henry J. Kaiser in order to produce steel plates for the Liberty ships he was building in the Los Angeles area to supply to the Pacific war effort. Its coking coal came from mines purchased or established in Utah, Colorado, and New Mexico, and its iron ore came from Eagle Mountain in California. By 1945, it was still a small plant with a capacity of 380,000 tons of iron and 750,000 tons of steel. After the war, sheets and galvanized products were added to its plate operations; and with four blast furnaces and an open-hearth steelmaking shop, it could produce over three million tons a year. With its basic oxygen converter added in 1958, Kaiser Steel became one of the first major U.S. companies to have that advanced steelmaking technology.

Kaiser Steel was always a bit of a maverick among basic steel companies perhaps because of its continued dominance by the Kaiser family who were without steel traditions. For example, Kaiser Steel broke with the industrywide solid front in the long 1959 steel strike and settled separately while also instituting profit-sharing. This same maverick stance was also to prevail in its relationships with and animosity toward the United Steel Workers of America with whom it has continuously fought with over, and eventually winning, the loyalty of its local workforce. Never one to be known as a big money maker, Kaiser Steel did reasonably well until the over-capacity era of the 1970s. Nevertheless, it operated in a chaos of ownership neglect with seven presidents in its last nine years of operation. The debt burden occasioned by the late addition of a new basic oxygen shop and a continuous slab caster along with the high cost of antipollution compliance occasioned by its location in the Los Angeles basin resulted in serious losses for the company. Its predominate family ownership saw no reason to risk capital from other enterprises and simply closed the plant in 1983 while also looking, in the meantime, for opportunities to sell off the assets and abandon the steel business.[2]

### The Prime Mover—Kawasaki Steel

Kawasaki Shipbuilding was founded in 1878 and grew rapidly during the wars with both China and Russia at the turn of the century. With domestic railroad building also creating high demand for steel, Kawasaki, in 1906, decided to build its own steel works which were enlarged substantially with the worldwide steel demands of World War I.

After World War I, Kawasaki Shipbuilding, having diversified into shipping, aircraft, and steelmaking, changed its name to Kawasaki Heavy Industries. By 1941, Kawasaki's steel operation had become Japan's third largest, producing 5 percent of Japan's total steel products. While not yet integrated, Kawasaki had two works; one was a basic mill in Hyogo and the other was a specialty mill in Fukiai which manfactured thin plates and sheets.

In 1950, Kawasaki Steel became independent from Kawasaki Heavy Industries and determined to enter integrated steel production with the construction of the massive ton Chiba Works. However, its late entry as the last company among Japan's six largest steel producers to introduce the basic oxygen furnace placed it at a disadvantage as the 1970s oil crisis, followed by worldwide steel over-capacity, emerged.

Meanwhile, however, during the high-demand era of the late 1960s, Kawasaki had gone looking for additional sources of semi-finished steel to feed its specialty finishing mills which were found in Brazil. In the early 1970s, Kawasaki and the Italian steel firm, Italsider, entered into a joint venture with Brazilian iron ore producer, CVRD, and private Brazilian steel firm, Companha Siderurgica de Tubarao (CST), in order to build a three-million-ton slab mill which would produce steel slab that the Italians and Japanese would then ship to their own countries for finishing. However, by the time the Brazilian slab mill was completed and ready for production, the world of steel had changed radically. By the early 1980s, both Italsider and Kawasaki had surplus capacity in their own basic facilities and had no outlet for the Brazilian semi-finished imports. Suffering decreased sales while confronted with high fixed debt commitments, Kawasaki was forced to sell off $20.8 million in assets. Without export, Rio Doce and Siderurgica could not operate at an economic level.

Kaiser Steel had been supplementing its high-cost, low-quality Eagle Mountain iron ore with imported Brazilian pellets and, at the time of the shutdown, owed money to Rio Doce. The answer to this problem was obvious—a simultaneous entry into the world's largest steel market and an outlet for surplus slab made with surplus ore.

### Investment, Facilities, Products, and Markets

With discussions already under way before the closure, the Kaiser plant was down only nine months between owners. Initially, an American steel warehouser, fabricator, and pipe mill operator, Mike Wilkinson, a friend of the Kaiser family who had always wanted to run a steel mill, facilitated the sale by buying 50 percent for himself with Kawasaki and Rio Doce each purchasing 25 percent. However, it soon became apparent that the Wilkinson pockets were not deep enough for a long-term commitment to the high-capital, low-return industry, and the two international partners bought him out. After borrowing $120 million from banks to buy the mill, another $150 million was required to modernize it and more would be required later.

Since the primary purpose for buying the mill was to provide an outlet for Brazilian slab, the coke ovens and blast furnace were of no use; they were not part of the purchase and were scrapped by the previous owners. Kaiser Steel had already replaced its open hearths with basic oxygen furnaces, which were also unneeded by the new owners. One basic oxygen furnace and one continuous caster were sold to the Chinese, who dismantled and shipped them to China. Kaiser Steel had paid $280 million for those facilities, but California Steel had no use for them and sold them to the Chinese for $17 million. Another basic oxygen furnace and continuous caster along with the plate mill was sold to Oregon Steel for operation on its Fontana property.

At the end of 1994, the California Steel plant consisted of a 1.8-million-tons-per-year capacity eighty-six-inch hot-strip mill, two pickling lines with another under construction capable of replacing those two, a 132,000-ton capacity cold-rolled sheet and coil facility, a 15,000-tons-per-month galvanizing capacity, and a pipe mill at Napa, California, using Fontana output. The total complex provided a rated, but often exceeded, one-million-ton hot-rolled sheet, cold-rolled sheet, and galvanized steel output. The Brazilian slab mill supplies about 60 percent of the raw material with the rest purchased from the United States and other international sources to maintain adequate inventory.

CSI specializes in supplying the California construction industry with sheet steel and pipe, leaving the canning industry to UPI, the Northern California joint venture between USX and Korea's POSCO, which is discussed later in this chapter. The two compete to some degree in galvanized products but, for the most part, remain with their respective specialties since neither market is large enough to provide adequate economies of scale to two producers. The West Coast's only automobile assembly plant, NUMMI (jointly owned by General Motors and Toyota), obtains its steel from eastern plants because the western market is not large enough to justify such a specialty in the west.

Kaiser Steel had been a major plate producer; but, initially, the Brazilian slab mill could not supply sufficient raw material for both sheet and plate. Since the western plate market was only 800,000 tons per year, the decision was made to concentrate on the sheet market. That decision was confirmed when Kaiser Steel's old competitor, Utah's Geneva Steel plant, was purchased by new owners and continued to supply plate throughout the west.

CSI exports only to Canada and Mexico and competes with Korea and Japan in its own West Coast market. After producing approximately one million tons in each of 1991 and 1992, 1.2 million tons were shipped in 1993, with similar expectations for 1994. Fifty percent of this shipment was hot-rolled sheet, 25 percent was galvanized sheet, and the remaining percentage was in pipe and other products. Management was convinced it could raise its output to 1.6 million tons without further facilities by adding shifts and increasing employment.

### Organizational Structure

The international ownership is accommodated by a Brazilian chairman of the board of directors who is a resident of New York, a Japanese president, and a Brazilian senior vice president on-site in California. Of the three executive vice presidents, the assistant to the president is Japanese, the vice president of operations and commercial activites is an American who has had long experience with Kaiser Steel, and the vice president of finance and purchasing is Brazilian. At a subordinate level there is a Brazilian vice president of finance, an American vice president of production and sales, and an American vice president of human resources. There are two other Brazilians in top managment, and all the rest are American.

The president, Takehito Haga, is a long-term official of Kawasaki Steel and was appointed by them. He will be rotated out, as all the Japanese will be periodically. The Brazilians represent Rio Doce in the same way, despite the fact that they do not bring steelmaking experience with them. There are no Japanese engineers or supervisors and no Brazilians below the vice president level. The Americans are simply employees of CSI with no other institutional representation; and approximately equal numbers of them have been there with Kaiser Steel, while others have been hired since. The Americans find the constant international rotation troublesome because of the necessary familization time, but they recognize this as a reality of international management.

### Profitability

The plant lost money in 1984 and 1985 as a consequence of start-up conditions but has made money every year since. Profits were down under recession conditions in 1990 and 1991 but have remained in the black and recovered since. Purchase of slabs from CSI's owners tends to stabilize its profits. As demand weakens, the price of the slabs falls, which helps maintain profitability. As the market goes up, the slab price goes up as well. On the other hand, demand fluctuates greatly because of ties to the construction industry, which tends to destabilize profits, since costs other than raw materials tend to continue despite production levels. The break-even point is about one million tons, which has been met or exceeded in every year since 1988.

Both parent companies play important roles in management. CSI sends monthly reports to them and parent company representatives visit their progeny periodically. They are interested in and aware of what is going on, but they rarely intervene. Since the Brazilians have less steel production experience than the Japanese, their oversight is less intense. They monitor the volume of raw materials, products, and profits but pay less attention to the operation of the plant.

The Americans in management, particularly those held over from Kaiser Steel, are a bit non-plused by the Japanese. Since Kaiser Steel was a stock company with substantial family ownership, an eye had to be kept on the stock market. Being owned by parent companies that are themselves privately held, along with the financing Japanese banks represented on the board of directors, the Japanese owner can take a longer view of profits. However, this may not continue given Japan's economic troubles as indicated by the National Steel and Armco experience.

### Cultural Differences

Since they are confronted with three national cultures—Japanese, Brazilian, and American—CSI officials believe they are creating their own unique culture. According to President Haga, "The Americans reach a goal very effectively and fast. The Japanese try to make precise and long-range plans. And Brazilians always look for the opportunity for change or innovation."[3] There have been some culture conflicts, but not serious ones. Management meets every Monday morning in order to discuss operations, plans, and issues. Participants consider these discussions to be very frank. Given a four-year rotation, some of the Japanese have language problems, whereas the Brazilians have greater English facility. It is reported that many American managers and employees have difficulty being led and supervised by the Japanese, and the latter are also often reluctant to "boss" the Americans.[4]

All the other joint ventures among our case studies include Americans who are involved as representatives of their firms. The combination of joint ownership on American soil tends to leave the Americans in dominant positions. However, at CSI, a Japanese man is the president and the more dominant figure. Americans are the majority among the managers, but their tendencies are subordinated by Japanese policy. Since the Brazilian managers lack steel backgrounds, their role is less notable. Hence, Japanese commitments to long-term planning, preventive maintenance, plant cleanliness, quality consciousness, employment stability, and longer-range profitability are all evident at CSI.

The first of the plant's three-year plans was developed almost exclusively by the Japanese; however, President Haga has been committed to involving the entire management in this process. The plant is now entering its third planning round, and full involvement is required. The Japanese management has a mission statement which it takes seriously as well as a strategy for pursuing it. The Americans are becoming more deeply involved and committed. As some of them point out, they have an even greater interest in keeping the company alive since they cannot rotate back to Japan or Brazil to the parent companies.

The Americans have found it much easier to work with the Japanese than the Brazilians because the Japanese are steel-operating people with whom they can speak the same industrial language. Initially, the Brazilians did not have steel experience. With the passage of time, however, they have learned, and some of those with experience at the Brazilian slab mill are even beginning to participate in operating decisions. The slabmaker, CST, is partially owned by Rio Doce, but until a year ago, the only relationship CST had with CSI was as a supplier. According to the shareholders' agreement, Rio Doce is responsible for purchasing slabs but not for producing steel.

The Japanese president can set policy, but the reality is that it must be carried out by Americans. Hence, the insistence on product quality is Japanese but the quality control function is, of necessity, headed by and carried out by Americans, as is every other functional area. By the same token, Japanese-style preventive maintenance is assiduously pursued by American executives supervising American workers.

### Labor Relations

The employee relations and human resource functions follow primarily an American model. Of the 900 total employees, there are seven Japanese and six Brazilians. The Japanese all get along well, though they have some language problems. The plants began TQM in 1992 with the assistance of the Juran Institute as well as their goal of improving product quality and their service to customers. Since the success of TQM depends on teamwork and mutual respect among and between management and employees, the plant has in place what it calls thinking groups, which are equivalent to quality circles seen elsewhere.

The steelworkers' union had been decertified under Kaiser Steel and the plant has remained non-union. A 1992 USWA organizing attempt was subsequently rejected by an overwhelmingly negative vote, with many of the employees allegedly blaming the union for some of the high costs and other problems which led to the plant's closing in the first place. Kaiser Steel had a profit-sharing plan which had not been very successful because, essentially, profits had been scarce. So far, the CSI profit-sharing program has been more popular, since there have been profits during all but the first two years, as well as the fact that 1993 and 1994 were particularly profitable years. Eight percent of income goes to equalized profit-sharing for all employees. There is no incentive system for non-exempt employees but there is for exempt employees. There are also bonuses for management, as well as profit-sharing. This latter fact is a delicate matter which is controversial among the employees.

Kaiser Steel had more than 3,000 employees, many of whom were old enough to retire. Probably one-half of the current 900 employees at the plant were formerly with Kaiser Steel. The new company did not deliberately set out to rehire Kaiser employees; however, they were the only experienced steelworkers in the community and the most likely to apply.

CSI has never laid off any of its permanent employees. The plant does hire some temporary workers for unskilled labor jobs who can be laid off when demand is slack. With 900 regular employees, it may have 50 to 60 temporary employees at any time. They are hired through temporary agencies and laid off when not needed; however, after gaining experience, they are often hired as permanent replacements for those who quit or retire.

Contrary to the Japanese image, the plant has no systematic training program. People are trained only as needed for special or new jobs. The plant has started a TQM training program for its managers but has done little skill training, perhaps because the plant had a ready supply of skilled steelworkers available when it started. That will change with time as replacement of incumbents becomes necessary.

### Operations Management

According to President Haga and Executive Vice President for Operations/Commercial Jim Declusin, the plant functions at the operating level with few differences from what would exist if it were totally American owned; essentially, the planning is Japanese, and the operation is American. The Japanese are more committed to preventive maintenance, whereas American plants tend to have more downtime than Japanese. The Japanese also tend to keep more careful data about both production and maintenance and, in turn, follow it more closely. Both of these characteristics prevail at CSI. Brazilian managers also participate in planning; in fact, they prepared the document which the plant is now using as a planning guide. The finance people are particularly involved in planning, with the Brazilians playing a larger role there than in other functions. The Japanese are much more experienced and committed to planning and, therefore, lead in this function. What exists at CSI, therefore, is the finishing end of a traditional American steel mill which is financed with bank loans combining patient capital with a high-debt load, planned by Japanese, including scheduled maintenance, staffed by Americans, protected, at least so far, by Japanese job security, and which is operated on a day-to-day basis like any other American mill except for the absence of any union. In an American context, the plant has more the atmosphere of a mini-mill than either an integrated plant or a finishing mill. So far, it has been extremely successful, having survived a mild recession without loss. This expatriate industrial marriage has lived on American soil with the support of indigenous servants and, so far, appears pleasing to all concerned.

## A U.S./KOREAN JOINT VENTURE

Though the only one of our joint venture cases not involving a Japanese partner, the experience of POSCO and U.S. Steel is much like that of the others.[5] This UPI marriage of convenience has experienced crosscultural strains; however, neither partner has regrets and the liaison promises permanence.

### The History of POSCO

The lack of raw materials and investment capital did not deter the South Korean government from opting for self-sufficiency in steel production at the end of the 1960s. Dependent upon Japanese war reparations for capital and Japanese consultants from Nippon Steel and NKK for technical advice, Pohang Iron and Steel Corporation was organized in 1968. With the construction of an integrated mill in 1970 producing 1.03 million tons of steel in 1973 and expanded in four phases to a final capacity of 9.1 million tons by 1983, POSCO's original mill became one of the largest single steel plants in the world. Seventy percent ownership was held by the Korean government and its Development Bank, and the remainder was held by private interests. Korean domestic demand for steel continued to escalate which motivated the expansion. However, in the early 1980s, the Korean government, primarily for job creation reasons, resolved to build another 11.4-ton mill at Kwangyang. Despite the protest of other steel-producing countries that it was aggravating an already existing excess capacity, the Korean government knew it would have to find foreign outlets for steel production in the short run. At the same time, the Korean government expected world economic growth to catch up to steel capacity in the future.

### UPI History

Meanwhile, a historic U.S. Steel finishing mill at Pittsburg, California, was in trouble. This mill was founded in 1910 as the Columbia Steel Company and consisted of one building along with a single 150-ton open hearth to furnish steel castings for the dredging, lumber, and shipping industries. The city's residents were so pleased with this venture that they voted to change the name of the city from Black Diamond to Pittsburg after the eastern birthplace of the steel industry (the "h" in Pittsburgh was dropped for simplified spelling).

In the 1920s, the plant expanded to include the west's first nail mill and, later, expanded even more to include the first hot-dip tin mill west of the Mississippi River. Columbia Steel merged with U.S. Steel Corporation in 1930 and, over the next two decades, facilities and equipment were added to help supply major public works projects. The most notable of these projects were the San Francisco/Oakland Bay Bridge construction and the attempt to meet the demand for steel products during World War II.

At war's end, the Pittsburg (California) finishing mill, often called Pitt–Cal to differentiate its location, was married to the basic integrated Geneva Works in Utah, which was created during the war to supply plates for warships but was now without a market. The new Columbia–Geneva Steel Division of U.S. Steel

Corporation would produce steel coils in Utah, sending some 70 percent of its output—an average of one million tons per year—to the Pitt–Cal plant to be finished into annealed, galvanized, and tinned products for West Coast canning, appliance, and other industries.

During the 1950s and 1960s, modern continuous sheet and tin mills, cold-rolling mills, electrolytic tinning, and continuous coating and annealing lines were added to the Pittsburg steel finishing facility. In the 1970s and early 1980s, computer controls and further modifications raised the galvanizing and tinning lines to state-of-the-art standards. The plant also narrowed its focus to finishing sheet and tin products only.

However, by the mid-1980s, both Pitt–Cal and Geneva were in trouble, as was much of the rest of the American steel industry. With the advent of new production processes, increased use of alternative materials like aluminum and plastic, and growing competition from foreign imports, both companies were becoming outmoded. Environmental protection requirements were imposing major costs on them. Further, Geneva had been built to produce wide-width steel plates for ships and, despite considerable modification, could not exercise the quality control required for narrower and thinner products. Begun in 1942 and coming on line in 1944, what had once been the most modern integrated mill in the nation was now forty years older and fundamentally unchanged. It was one of the country's lowest cost mills, but its product had never been adequate for Pitt–Cal's needs and was deteriorating as quality demands rose. Pitt–Cal's average yield from any given Geneva product had fallen to 58 percent. The rest of any average shipment was rejected for defects which made the products not appropriate for finishing mill purposes.

U.S. Steel estimated that $1.5 billion would be required to bring Geneva to competitive standards, and it was neither willing nor able to come up with that kind of money. More money had been invested in upgrading the Pittsburg plant, but the quality demands of West Coast canners, appliance manufactures, and other potential Pittsburg customers were rising as their appetites were whetted by the better Japanese and Korean steels which were flooding the West Coast at no higher prices, despite the voluntary restraint agreements. Another $500 million was needed to bring Pitt–Cal up to rising standards. Rather than risk $2 billion, which it did not have anyway, USX resolved to withdraw from the west and leave that market to the orientals.

**Motivation for the Joint Venture**

With these issues hanging over their heads, USX Chairman David Roderick and POSCO Chairman Tae-Joon Park met at an International Iron and Steel Conference in April 1984. The solution was clear. A joint venture would be formed.[6] POSCO would put up the $500 million to modernize Pitt–Cal. As soon as the existing voluntary restraint agreement (VRA) expired in 1987, POSCO, from its new Kwangyang plant, would supply high quality coils to replace Geneva's deficient products. Pittsburg workers would keep their jobs and Kwangyang workers would get new ones. POSCO would gain a new customer and access to the U.S. market which would eliminate much of the animosity that other exporters to the United States were generating. USX would have half ownership of a refurbished finishing mill without cost. Everyone should have been happy except for those at the Geneva plant in Utah. As it turned out, however, to the surprise of everyone including themselves, a group with ties in Utah bought the Geneva mill even before the VRA expiration and turned it into a viable facility. There were no losers.

The $500 million modernization of Pitt–Cal, which was by then renamed UPI for U.S. Steel and POSCO Incorporated, was completed in 1989 and became one of the largest industrial investments in the state in years. With POSCO providing the upstream technology and a reliable source of high quality hot-rolled steel coils, the refurbished UPI plant turning out high quality product, and U.S. Steel offering an extensive distribution system to serve domestic and world customers, it was understandable that few outside of Utah objected.

The Pittsburg mill is well located for an international co-venture since it is located on water deep enough for ocean-going vessels. At this time, four vessels were used, each of which took fifteen days for a one-way trip between Korea and Pittsburg with each ship carrying approximately 1,350 coils. The ships would then load iron ore, either in Canada or southern California, for the return voyage. Additional coils are shipped in from the Gary and the Fairless mills of USX in a 70–30 ratio.

The UPI products were tinned, galvanized, and annealed steel. The tinning and galvanizing mills were old but had been refurbished. The annealing line was completely new. Bill Pence, a Pitt–Cal marketing manager who retained that position at UPI despite his origins at Geneva Steel, described the joint venture as a "marriage made in heaven." The divine blessing, in his mind, was the fact that the Koreans, who had already spent $20 billion to build in Kwangyang, were now willing to spend $500 million more to upgrade the Pittsburg mill without costing U.S. Steel a dime.

Actually, not costing U.S. Steel a dime was not POSCO's original intent. They believed that their $500 million investment should buy them a half interest in the Pittsburg plant and, thereafter, each joint venturer would put up half of any additional investment needed. USX refused to do so, however, and all subsequent improvements have been paid for out of UPI revenues.

### Markets Served by UPI

Before the joint venture, the plant shipped to more than 300 customers in the thirteen western United States.[7] While the number of tons shipped has not decreased, the number of customers has declined; some customers have gone out of business and others have merged. The number of truck shipments is over 200 daily, and new records are being established monthly for rail shipments.

One of the advantages of the joint venture as perceived by POSCO was that USX could provide an extensive marketing and distribution network to serve domestic and worldwide customers. However, UPI has not been able to penetrate markets in the eastern United States because of transportation costs, domestic competition, and other factors. UPI has formed a new export division and has shipped customer orders to India, Singapore, Australia, Mexico, Guatemala, and other Pacific Rim countries. Exports rose from nothing in 1986 to over 50,000 tons in 1992 and plunged to 16,000 tons in 1993.

There is one new area in marketing about which UPI marketers are excited and which they consider to have a great future. This is the new expanding market for light-gauge steel used in residential home building. It is now being used and promoted in earthquake and hurricane areas and is perceived as a growth market. Expansion in this product will put UPI in direct competition with CSI, which has preferred to leave tin plate to UPI and keep the construction market to itself.

Close-in delivery helps keep UPI strong in tin plate. The plant ships about three-quarters of its tin plate products to within fifty-five miles of the plant, which serves canning companies in northern California. The company also has been successful in shipping cold-rolled steel to makers of office furniture. It holds about three-quarters of the market for cold-rolled steel in the western United States and under 20 percent of the market nationally.

### Commitment to Product Quality

Reflecting on the availability of steel from Gary, Fairfield, and POSCO, operators often declare their preference for hot bands from POSCO to run on the new equipment because the Korean quality is much better than the domestic supply. Since the new equipment was installed for the modernization efforts in 1989, the plant has increased its productivity and yield substantially. Man-hours per ton fell from 1.8 in 1986 to 1.2 in 1993. The plant suffered high rejection rates in tin plate prior to the modernization, but since has cut its yield loss in half. The new equipment was designed to run at optimum levels and in order to do so, the quality of hot bands must be excellent. POSCO is able to guarantee the supply of high quality feed stock on a consistent basis.

Before the Korean involvement, employee commitment to maximizing their incentive earnings sometimes outweighed their concern for product quality. However, there is a new awareness of customer needs, and employees have raised their standards. There are also more statistical quality controls in place, courtesy of the Korean influence, and progress continues in UPI's goal of becoming a registered ISO 9002 company. The International Organization of Standards governs the quality of products between countries and recognizes companies that have standards of practice and process controls in place to ensure that they are capable of making a quality product. Employee empowerment is also being implemented. Employees now can stop lines and make suggestions in order to improve quality. Visits to customers by production and maintenance employees has also improved understanding and communication.

### Investment and Technology

The joint venture was planned on the premise that there would be a high return on the investment after modernization of the facilities. However, even taking into account large loan payments, the return has not materialized because of the relatively low price of steel in the marketplace. Profits during 1986–1988 were followed by losses during 1989–1992. Had it not been for the return to profitability thereafter, the high fixed-debt costs might have produced serious doubts. Downward price trends have necessitated changes in the business plan to ensure a return on the investment. The focus is now on cost reduction wherever possible in order to ensure a low break-even point. It is a perception of at least one manager at UPI that neither party would have entered into the joint venture if the downward trend in steel pricing had been perceived. The cause, of course, has been excess capacity and oversupply throughout the world. However, since the alternative for the Koreans would have been to not build Kwangyang and the continuance of Pitt–Cal was accomplished at no cost to USX, the speculation seems unlikely.

At the time of the joint venture, advanced technologies were available, but USX did not consider it in its best interest to reinvest in its western facilities. Instead, these facilities were allowed to deteriorate. Staying competitive and improving market participation required a major modernization of plant facilities and was beyond any amount USX was willing or able to invest. The joint venture raised the capital necessary for the $450 million modernization program. The new technology gave the plant the unsurpassed ability to provide the best quality steel for its extensive customer base and provided competitive products for new markets.

With a projected annual capacity of 1.4 million tons, the revitalized mill has been able not only to manufacture an improved grade of tin plate for food and motor oil cans, but has also been able to produce high-quality cold-rolled sheets for office furniture and household appliances as well as galvanized sheets for use in highway culverts, gutters, and other construction products.

Following that $450 million investment in modernization, the UPI plant has become one of the world's leaders in the production of high-quality flat-rolled steel. The company spends about $10 million annually on continuous upgrades to plant operations. With all of that, its profitability remains in the 3 to 4 percent range which is typical of basic steel.

### Organizational Structure

The organizational structure and nationality mix has changed significantly since the joint venture began. Before the joint venture, the top operating managers reported to the plant manager while the heads of staff departments reported to managers at Geneva Works. With the establishment of the joint venture, the new organization included a president, executive vice president, and vice presidents for operating, commercial, employee relations, and finance. Positions were also established for a comptroller and an attorney. Only two executive positions are held by Koreans; these are the executive vice president and the comptroller. Below that level, the manager of traffic is Korean as well as a manager of sales. The only other Koreans in the firm are ten to fifteen engineers. As of 1994, Leonard H. Chuderwicz from U.S. Steel was President with H.B. Ko of POSCO as Executive Vice President pending his imminent rotation to Korea.

The Korean managers and engineers have, at times, had feelings of isolation and have had difficulty assimilating into the American culture. This problem was often pronounced at the beginning of the joint venture and still exists to a lesser degree. The more recent POSCO arrivals seem more adaptable than the earlier group. They are younger, less shy, and seem more willing to speak up at meetings.

During the early stages of the joint venture, there was, in some departments (e.g., Purchasing), a need to have people added to the department because of the large volume of work during the modernization. However, in other departments, there was a trend toward "empire and power building"; and there is now an effort to flatten the organization to reduce costs.

Current managers attribute this expansion and necessary contraction to careless management in the early stages. Unfortunately, some of those who will be squeezed out of management jobs were promoted just three years ago from union positions where they would have been protected. In other areas, heads of departments established new non-bargaining unit jobs just a few years ago which, in a long-range view, should not have been created. Now when those positions are eliminated, employees with many years of service may lose their jobs.

Those losing out complain that if the departments had been managed well and kept lean prior to 1993, there would have been no need to reduce staff. In some cases, they complain that the reductions are merely devices for management to substantiate their bonus plan for the year. The reductions in many cases also turn out to be more apparent than real. Contract employees have been hired to replace the eliminated positions, sometimes on a continuing forty-hour week basis. Being management and white collar rather than bargaining unit employees, those displaced have no protection. The displaced complain that retention choices have been made politically rather than objectively on the basis of productivity with hard working but non-political employees suffering. It has been charged that instead of recognizing loyalty and retraining employees for other positions, those positions have been filled from outside or with less loyal and competent people. How true these perceptions are is difficult to assess, but the perceptions foster distrust and low morale.

### Cultural Differences

Profound differences between American and Korean management styles were noted immediately following the joint venture. Not that they have threatened the success of the firm, but the cultural differences posed definite interpersonal problems. One contrast has been the unwillingness of the Korean managers at all levels to speak up and disagree. They wanted harmony and balance without confrontation. An example of this was the American president's decision to take all members of the executive staff and their wives for a long weekend at a San Diego spa and golf course at company expense. The trip was planned and discussed far in advance at staff meetings with the Korean managers in attendance. Only when the trip was imminent did they finally speak up and express the opinion that company funds should not be used for that purpose. At that point, it was extremely difficult to cancel reservations and obtain refunds for deposits that were made.

Another example was the slowness in reaching decisions on even simple matters, making it difficult for Americans to work for Korean managers. One American manager complained that he often wanted to proceed with a necessary task, but would be told by his Korean superior to wait until the decision was discussed and a concensus reached with all of the other Korean staff.

Adherence to lifetime employment was another early issue. At the beginning of the joint venture, Korean managers promoted this concept, while the Americans argued for layoffs whenever redundancies appeared, even for short periods. A compromise was reached with a smaller percentage being laid off than the Americans felt necessary but more than the Koreans felt comfortable with.

One reason cultural differences were not resolved at the beginning of the joint venture was that the initial president and executive vice president were very strong leaders with very different ways of approaching issues. While the Korean executive vice president often disagreed with policy directions, he did not consider it his prerogative to directly confront the president of the company. When he finally did disagree, it was a real showdown. Their respective roles of joint leadership, while representing diverse organizations, necessitated strong personalities; however, there was a definite personality clash between the two leaders. One of the main reasons for the clash was the disrespect Koreans felt the Americans had for their culture in particular and minorities in general. That initial executive vice president did not accept the theory that one needed to be sensitive to cultural differences in order to improve communication. On the other hand, the American managers viewed him as a very traditional Korean who, at times, could be very inflexible.

As a result, the Manager for Community and Cultural Relations began a "Building Cultural Awareness Program." Speakers discussed the Korean family, management styles, traditions, and so forth in order to increase American understanding of Korean culture. Unfortunately, after the executive vice president was transferred, attendance dwindled and management would not mandate attendance, so the program was cancelled.

There is now a new approach by the legal staff to teach a program entitled "Diversity: The Law." However, diversity is taught from a legal standpoint and only touches on cultural understanding peripherally. Korean managers are concerned that while the program may lessen harassment, it still lacks an in-depth approach to cultural understanding and will not help communication and understanding between ethnic groups. In fact, the program may make people a little afraid to speak openly to each other for fear of a legal challenge.

Thus, cultural and national policy differences have presented a major problem for the key leaders of the joint venture. With natural allegiance to the partner that transferred them to the venture, each national group must adhere to the directions given them. The facility cannot be managed strictly on its own with no concern for the views of the parent partners. This was especially difficult during a 1994 international trade case where the partners had different objectives. This unacknowledged conflict presented an extremely stressful situation for all top management involved in strategic planning and also limited honest communication among them. Cultural differences also presented problems because in one culture, some issues can be far more important than in the other, and no compromise can be reached. There is still a tendency for the POSCO employees to want harmony and balance and they continue to be reluctant to cause dissension. However, this problem is not as pronounced as it was at the beginning of the venture. With time, understanding will occur even if the differences do not go away.

### Differences in Management Styles

Differences in management styles between Americans and Koreans still prevail, although they also are diminishing with time and experience. One reason for this is that the new arrivals from POSCO are younger and more flexible. However, there is still a distrust of the Americans by some of the Koreans who, in light of American stereotypes of Orientals, think that Americans do not really say what they mean. Some Koreans in the company are also convinced that there is an attitude of superiority by Americans who are not really interested in learning about other cultures and how these cultural differences can impede communication. There is no way to assess the accuracy of this perception, but the perception is itself a problem. Minorities of any origin tend to have a greater sensitivity to this issue, and their reactions, cooperation, and productivity are likely to be tempered. Addressing the issue through the legal system may cut back on harassment, but it does not improve communication.

The Korean staff tend to work longer hours and on Saturdays, and most Americans do not. They have extreme pride in the venture and are proud to be a part of UPI. American employees in UPI often categorize the POSCO employees as "missionaries" for POSCO because they sacrifice so much of their personal and professional lives to come to the United States and enter a foreign culture. The fact that the assignment is temporary probably contributes to their willingness to work more intensively, as does their lack of community involvement. The American managers generally do not work as many hours and do not exhibit the same sense of personal pride in the joint venture. The more recent POSCO arrivals seem to take the cultural differences in stride and are, again, more flexible. Their families, however, especially their wives, still seem to be very shy.

### Personnel Policies

Personnel policies did not change significantly after the joint venture. The initial vice president for employee relations was already at the plant and continued existing practices. The USWA bargaining unit also remained the same with the same contract in force, even though its provisions had changed substantially in subsequent years. The Korean staff was initially overlooked for merit increases because they were reluctant to voice their opinions on salary matters. No one was willing to bring the oversight to the attention of the personnel staff until a replacement vice president for employee relations discovered the oversight and began merit increases for the POSCO employees.

In addition, American employees assisting the POSCO staff were viewed with suspicion by some of their peers among the American staff. This feeling still persists against those Americans who worked very hard to help UPI win the trade case. Only with the intervention of key POSCO executives were some of these employee positions protected from elimination.

It is interesting to note that late in 1994 the Employee Relations Department was eliminated and its functions were divided among other departments. Payroll and benefits became the responsibility of Employee Financial Services as did the formalities of hiring after recruitment and selection by the Learning Center. Labor relations is now directed by an attorney and by the vice president of operations. The passage of time has been inadequate in determining the results of these changes.

A significant change of Korean origin was the concept of incorporating maintenance into each operating department instead of having a separate maintenance department. This facilitated the concept of scheduled preventive maintenance rather than waiting to fix breakdowns on an emergency basis. Another change was the combination of sheet and tin operations under one division manager instead of each having its own manager which, hopefully, would improve coordination.

## Employee Selection Criteria and Training Policies

Before the joint venture, U.S. Steel promoted upward within the organization except at the plant manager level. The latter was typically transferred to the facility from other U.S. Steel plants. This policy provided positive motivation because employees knew that if they generally worked hard, they had an opportunity to rise in the organization. However, few promotional opportunities were available for women and minorities because of the "old boy network" which generally promoted Caucasian managers from the "in" group. In the labor ranks, upward mobility occurred through the union process of bidding for jobs to a higher job class; however, nonexempts and management were at the mercy of the key players. The few recent hiring situations have been restricted to the engineering area, some of whom have been talented minorities. If they remain to rise into executive positions, it will be because of a combination of equal employment opportunity legislation and the changed culture it has gradually produced as much as the international management.

After employing more than 5,000 people during its heyday in the 1940s, the UPI plant reopened with only 90 people and currently employs about 990. Prior to the joint venture, little training was done at the facility and no learning programs existed.

POSCO was the partner responsible for establishing a learning program. An old building was remodeled for this purpose and is now recognized as a model for the rest of the industry. Learning program courses are now open to all employees. Many courses are directly related to job requirements, such as management training, business planning, trade and craft training, statistical process control, safety programs, and computer training. Other courses are offered through California State University–Hayward, leading toward B.A. and M.B.A. degrees. The latter program has also been recognized as a model for the industry.

As a result of enhanced training, which is another Korean contribution, employees feel more a part of the organization, are more committed and empowered, and they have a loyalty and pride which appear to be developing. The employees are learning the importance of keeping abreast of new knowledge and technology in order to remain competitive. The Learning Program has also been very important in breaking down the adversarial barriers between the union and management ranks.

### Labor Representation

Two labor union locals existed before the joint venture and continue to represent the employees. Local 1440 of the United Steel Workers of America represents production and maintenance while Local 2571 represents clerical and technical employees, together comprising 800 of the total 990 employees at the facility.

The intended joint venture was not discussed in advance with the union, though the existing labor agreement required any subsequent owner to recognize the union and become party to the agreement. However, union officials have been included in decision making since the joint venture and a "Union-Management Participation Team" has been established which meets twice monthly. Representatives include vice presidents and union officials.

The union officials have reacted favorably to the joint venture and prefer POSCO's policies over those of USX. The union has often expressed the desire for POSCO to purchase an additional percentage of the company so that POSCO would be in control. A few rank-and-file employees may harbor prejudices not evident among the majority; however, the general atmosphere is one of loyalty. The joint venture has broadened the perspective of many employees who express pride at being associated with a competitive world-class steel company. If cutbacks and layoffs were to occur, anti-Korean feelings might be manifest, but thus far the general feeling appears to be one of gratitude for the jobs saved.

### Pay Practices

Pay practices have also not changed appreciably as a result of the joint venture. A merit increase covering non-union employees is based on job performance and usually ranges from 4 to 6 percent. Promotions usually result in pay increases of 7 to 10 percent. The plan is not popular, however. Some, at least, see politics and favoritism as inherent in the merit system. What and how one person appraises another can be very different, depending on their perception and values. An individual who goes for four or five years without an increase because of low ratings is not likely to perceive objectivity in the rating process. Clerical employees also complain that their pay is below community norms, and women in management positions, though protected internally by equal pay laws, believe they are paid less than men in comparable positions in the community. All of these factors have affected employee motivation.

Probably because of distance from major steel centers, the international union has tended to allow greater departure of the UPI agreement and practices from national patterns. Nevertheless, the pay provisions for union employees have not departed significantly from industrywide patterns despite the break-up of industrywide bargaining which occurred simultaneously with the initiation of the joint venture. Preexisting incentive plans continue in certain departments in hopes of increasing motivation and productivity, as well as a bonus system which has also been added to the union contract.

Benefit practices have not changed significantly, and the employees have excellent health insurance, dental insurance, and investment and pension plans. The only significant change has been that they are now required to pay $40 monthly for their health benefits which was accepted without protest because of the awareness of the rising costs of health care. Union employees receive five weeks of vacation, but those outside the bargaining unit have been cut from five to four since the joint venture; however, the latter do have a more generous sick leave.

### Community's Response to Joint Venture

The community's response to the joint venture was, at first, one of alarm. The company had been in the city for over seventy years with three generations of families working at the plant. There was initial concern about being taken over by a foreign company. In addition, the company chose a non-union contractor for the modernization program, which was an unfavorable portent for union members. It has taken many years to recover from the ill will which developed over the non-union contractor issue. However, the community now supports the company and realizes that the joint venture saved the plant from closure. It has also become obvious that the company has no intention of importing Koreans for significant numbers of jobs.

No changes were made in the original joint venture agreement to accommodate community concerns. However, a subsequent community relations outreach program has generated community support. UPI established a manager of community and cultural relations and this individual represents the company throughout the community. The manager was active in the Steelworker Statue Committee which raised $45,000 for a bronze steelworker statue in the city and also represents the company on a new Chamber Industrial Advisory Panel established to foster a good working relationship between local industry and nearby residents. The company was also instrumental in the establishment of a sister-city relationship with the city of Pohang in Korea and actively participates in the annual Seafood Festival by coordinating the Korean Village. In addition, UPI also participates in the annual Multi-Cultural Expo by hosting a combination UPI/Korean culture booth and giving helium balloons to children.

Establishment of the community and cultural relations position along with its outreach program has created a favorable image and support of the company in the community. Response from the community was overwhelming when the community rose to support the company during the 1994 trade case. With protectionist forces seeking increased tariffs and the State of California pursuing further foreign investment and modernization of foreign-owned facilities, the Pittsburg community responded with over 7,000 signatures on petitions which favored the company. Tariffs on the incoming steel that kept their plant alive hardly seemed ethical to most UPI employees, especially after POSCO's substantial investment in their job future. The company also continues to support community activities such as the $8,000 given for fireworks for the Fourth of July celebration as well as generous contributions to local nonprofit organizations.

### Future of the Joint Venture

The joint venture has, of course, been subject to many second thoughts. Had either the continued downward trend in steel pricing or the emergence of alternative far eastern outlets for the Kwangyang output been known in advance, the partnership might never have occurred. In retrospect, both partners would probably agree that 50–50 ownership was not a wise choice, though they would probably not agree as to which should have taken the controlling interest. As it is, decision making is difficult and strategic planning is hampered. Control of the hot band supply would also be more strictly addressed in a new joint venture agreement. However, there is no indication that either party would prefer divorce.

Differing policies of the base nations have also been sources of stress and strain. During the planning stages of the joint venture, Malcolm Baldridge, Secretary of Commerce, and Ambassador Clayton Yeutter, U.S. Trade Representative, both stated their support in a letter dated November 1985 and addressed to David Roderick, CEO and Chairman, Board of Directors for U.S. Steel Corporation (now USX Corporation). The Clinton administration has been critical of Japan but not of South Korea and has been a strong advocate of free trade through the North American Free Trade Agreement (NAFTA) and the General Agreement on Tariffs and Trade (GATT). The domestic steel industry has remained influential in Washington but is seldom clear as to where it wants to go concerning trade policy. New trade policies will, of course, affect the conduct and policies of this and other joint ventures. Like other industries, steel companies simultaneously want barriers removed from them and protection supplied from others. As long as suits are filed by domestic steel companies, there will always be exceptions to accepted trade rules. UPI must have lobbyists and representatives in Washington to counter others' efforts and voice their concerns.

It is impossible for outsiders to know what each parent would prefer for the joint venture. It can be assumed that neither parent would be willing to carry significant losses for long if they were to reoccur. The joint venture agreement provides for sale to the other partner if one prefers to exit the relationship. If neither wanted to make the purchase, the 50 percent ownership would be available for other investors. If POSCO withdrew from the venture, they would have no problem selling their hot bands which are produced at Kwangyang in the growing Asian and Pacific Rim marketplace. USX would also have no difficulty finding a guaranteed hot band supplier, including a revitalized Geneva Steel, but seems to be more disposed to abandoning the Western market. If USX withdrew from the venture and POSCO did not increase its ownership percentage, it would be faced with a whole new set of problems by dealing with an unfamiliar new owner. These problems could be better or worse than dealing with USX.

Joint ventures such as UPI (where one owner supplies the raw material) are very complicated because they involve trade issues and government regulations. Some POSCO principals suggest they would now advise against other such joint ventures if asked. UPI was the product of conditions at a particular time in trade relations. Had POSCO foreseen the unique difficulties in trade and the environmental concerns that the plant had to deal with, and had also seen the expansion in steel demand that would arise internally and on their own borders in the Far East, they probably would not have entered into the joint venture.

But once committed, Korean management strategy is long-term and goal oriented. The Koreans do not make quick decisions. POSCO has enormous pride in the UPI joint venture. In all probability, their patience will enable them to stay with the venture as long as persistent losses are not in prospect. In the meantime, the cultural strains have been irritants but not serious threats to the international marriage.

## INLAND AND NIPPON STEEL

Like two senior citizens engaging in romance at a care center, a marriage between two of the oldest steel companies in their respective countries seems somehow comforting; but the union seems most unique when they conceive and deliver a pair of extremely lively twins. That, in short, is the story of the emergence of I/N Tek and I/N Kote from the joint venture of Inland Steel and Nippon Steel.[8]

### Inland Steel History

Inland Steel Company is one of the oldest of the American steel companies having been established in 1893 to reroll scrap rails in Chicago Heights.[9] Its first steelmaking venture was begun in 1901 on the lake shore north of East Chicago and at its historic Indiana Harbor plant which was established in 1906.

Like the rest of the industry, Inland Steel experienced rapid growth after World War II. Its single plant ultimately grew to an ironmaking capacity in excess of 18.0 million tons and a steelmaking capacity of 9.3 million tons. This largest single plant in the nation mounted sufficient capacity to make Inland Steel the fourth largest steel producer in the nation and responsible for approximately 6 percent of total output. The most recent addition to that Indiana Harbor capacity was the cold reduction mill added in 1969. Inland Steel's location, which is five miles away from the nation's second largest steel mill, USX's Gary, Indiana, plant, largely accounts for the Chicago area's preeminence even over the Pittsburgh, Pennsylvania, industry in steelmaking. Inland Steel's growth has not been merely the result of overall market growth, but rests on a tradition of managerial excellence and careful cultivation of sales outlets. Historically and financially sound, the company has traditionally been a source of stable though cyclical employment, allowing for development of a strong sense of intergenerational family traditions among its workforce. Nevertheless, this historic bastion found itself in trouble during the 1980s, losing $500 million between 1982 and 1985 with the latter year's loss alone totaling $178 million. It was time for rethinking of roles.

### Nippon Steel's History

The Japanese government-owned Yawata Works was organized in 1886 and began steel production in 1901, the same year in which the U.S. Steel Corporation was founded. The plant was designed by German engineers, and its equipment was imported from Germany. By 1906, the steel output of the Yawata Works was over 90 percent of Japan's total domestic production. Privatized in 1934 as the Japan Steel Corporation, the Yawata Works continued to be the largest fully integrated iron and steel company in the nation. However, after World War II, Japan Steel was required by U.S. occupation forces to divide into two integrated firms, Yawata Steel and Fuji Steel, in 1950. Fuji Steel took over the Hirohata, Kamaishi, Wanishi, and Fuji Works while Yawata Steel inherited the Yawata Works. Then in 1968, in order to create a single corporation that would have enough power to control the Japanese domestic steel market, the two major plants, Yawata and Fuji, which were the two halves of the prewar Japan Steel, were merged as the Nippon Steel Corporation. At the time, its market share was 44.5 percent for pig iron and 35.4

cent for steel. Perhaps more important than its size was its reputation for technological excellence and product quality.

### Motivation for Joint Venture

Rather than diversify or abandon the steel industry in the midst of its early 1980s losses, Inland Steel resolved to plunge ahead. In 1983, Frank Luerssen, Inland Steel's Chairman, resolved to expand and upgrade its cold-rolled steel technology. Whereas American cold-rolled processes consisted of separate pickling, cold-rolling, annealing, and tandem milling steps with expensive handling and inventorying between each step, he was aware that Nippon Steel had been successful in uniting those steps into a continuous process in a new Hirohata facility added to its Yawata Works in 1982. Resolving to build the first continuous cold-roll mill in the Western hemisphere, Luersson sought consultation from Nippon in modeling that mill after Nippon's Mill. A small core development staff of Americans and Japanese spent most of the years between 1984 and 1987 shuttling between Indiana Harbor and Yawata to design the new facility, assisted in the United States by a consulting firm, United Engineers.

However, in mid-1985, Inland Steel realized it did not have the money necessary to fulfill its ambitions. It appeared that the new continuous cold-roll mill was going to cost at least $450 million; in fact, the final figure turned out to be $525 million. There was no way that either the stock market or the American banking industry was going to support such a development for a steel company, let alone one in the midst of substantial losses. Meanwhile, Nippon was confronted by the same over-capacity in its domestic market, enticements from Japanese transplants in the United States, and political opposition to its exports to the United States as was true of NKK, Kawasaki, and Kobe. Since Inland Steel was relying on Nippon not only for advice but was also planning to buy most of its new mill from the Japanese firm, the answer was obvious—another Japanese–American joint venture. Nippon bought a 13 percent stake in Inland. The two firms then divided the ownership of the new joint venture, to be known as I/N Tek, with 60 percent ownership for Inland and 40 percent for Nippon, the bulk of the initial $525 million being borrowed from Japanese banks.

In the planning process, the joint-venturers reached another momentous decision which would prove to have a major impact on the success of the new enterprise. Even though I/N Tek's raw materials would be hot-rolled coils from Inland's Indiana Harbor plant, the new continuous cold-roll mill would be located sixty miles away in a cornfield in New Carlisle, Indiana, a rural area with no previous steel industry exposure or experience. Perhaps Nippon was aware of and trying to avoid the problems National Steel and NKK as well as Armco and Kawasaki were having in trying to reform an existing culture on-site.

An agreement was reached, financial arrangements were made, ground-breaking occurred in August 1987, and the mill was ready for production by July 1989. Already the parties were so pleased with what they were creating that they had agreed on an addition of approximately equal size. Since the front end of the continuous mill was capable of producing 1.5 million tons and the back end only 1 million, there was room for diversion of the difference into more productive use. A separate 50–50 joint venture called I/N Kote would be built on the same site to galvanize part of Tek's output, bringing the total investment to $1.1 billion. Since Inland Steel already had its own galvanizing line, it was, in effect, creating competition for itself. However, the demand for galvanized sheet was rising rapidly and Indiana Harbor did not have the capacity to galvanize its own output and that of I/N Tek. Agreed upon in 1989, the I/N Kote facility came on-line in 1991.

There was a substantial difference in the relationships between Inland Steel and the new entities. I/N Tek was serving merely as a toll processing facility to Inland Steel, taking in coils from Indiana Harbor, processing them under continuous Inland Steel ownership, and warehousing the finished product for Inland to sell under its own name while charging a tonnage rate for the service. Kote would buy coils from Inland Steel and process them at its own expense for sale to its own customers. Though, in fact, Kote chose to employ Inland Steel as its exclusive sales agent since many of its major customers are Japanese transplants, it maintains its own sales staff as well to maintain direct contact with those customers. Nearly half of the customer base of Kote has been traditional Nippon customers. For instance, Toyota is the largest customer of Nippon Steel and has become a major customer of I/N Kote.

## Modus Operandi

The I/N Tek facility receives hot band coils from Indiana Harbor on a Japanese style, "just-in-time" basis, maintaining two days supply on railroad flat cars and unloading the coils directly into the plant for processing rather than unloading, storing, and rehandling them. The coils then unroll, with the tail of one automatically welded to the head of the next for continuity, into a pickling or descaling operation to prepare them for cold reduction in a tandem rolling operation. From there the Inland-owned rolls continue directly into a continuous annealing process line (CAPL), final pickling, a skinpass mill which imparts the final desired flatness, surface texture and mechanical properties, inspection, shearing and recoiling, and delivery to an automatic guided vehicle. The final finishing step consists of coil packaging, final inspection, and deposit in an automatic storage and retrieval system. The operation is entirely computer-controlled with staffing limited to oversight and inspection functions except for a minor amount of physical involvement in the coil packaging process. The one-million-ton-per-year facility reduces the time required to process a coil from twelve days to less than one hour.

The Tek operation is closely modeled after Nippon's Hirohata mill with some important exceptions, such as the pickling line, which does not exist at the Japanese counterpart. The joint U.S./Japanese planning team began with the Hirohata experience, considered the likely differences in the U.S. setting, and then modified as those differing expectations indicated. Since, from the beginning, it was assumed that a galvanizing line would eventually be added, a carousel reel was installed after the tandem rolling operation. That allowed material to be pulled off, either for inspection or, more important as it turned out, for diversion to the parallel I/N Kote building.

Transferred to I/N Kote by an automatic guided vehicle, I/N Tek coils enter a 500,000-ton Japanese-technology continuous hot-dip galvanizing line, which is also fed, to a lesser extent, by coils hauled in from the Indiana Harbor works. I/N Kote's capabilities also include a 363,000-ton German-designed electrolytic galvanizing line. Since the latter process requires fully annealed coils for processing, 80 percent of the coils electrolytically galvanized are hauled in from Indiana Harbor, the other 20 percent being coils which have completed the full I/N Tek process before being transferred to I/N Kote. The remaining Kote facilities beyond the two galvanizing lines consist of automatic storage and retrieving systems served by further automated guided vehicles, an inspection and side trimming line, a semiautomatic coil packaging line, and a waste treatment facility. Like Tek, Kote is fully automatic and computer-controlled with the exception of some aspects of coil packaging.

### Organizational Structure

The two companies share the same president, John Selky; the same controller, Dennis Freeland; and the same manager of human resources, John Nielsen. Reporting directly to them are those departments serving both companies such as accounting, human resources, information technology, purchasing, and plant services. Tek and Kote each have their own vice president of operations, manager of technology and engineering, and manager of sales administration. By agreement, the first two are always Japanese. Kote also draws its managers of customer services and sales and administration from Nippon, along with sales representatives to maintain communications with Japanese customers. Nippon also supplies a quality control person at the Indiana Harbor works with a staff to oversee the quality of the materials provided to Tek and Kote. Tek has a continuing but rotating complement of six Japanese engineers, while the twelve Japanese at Kote are in both engineering and sales.

A joint leadership team seeks to build a bond between the two separate entities which buy from and sell to each other but whose success or failure is so integrally linked. Because Nippon was accustomed to operating horizontally, integrating management systems, operations systems, and strategic systems in contrast to Inland Steel's traditional vertical hierarchical approach, the new joint venture management resolved in 1988 to employ a consultant to assist them in developing a team concept, first for I/N Tek and subsequently for I/N Kote.

Beginning at Tek, the consultant advised that "your teams ought to be responsible for your products from beginning to end." However, there was only one basic product, differing only by the length, width, or thickness of a particular order, and the continuous process meant that the entire operating complement was one team of thirty people on each shift. If support people were added—finance, customer service, equipment technology, and the like—every employee would be a member of the same team. To adhere to another team principle—that every employee should perceive the person immediately upstream and downstream in the production process as a member of the same team—the result at Tek became one team for each shift consisting of everyone on that shift. But thirty people were too many to function effectively as a team, leading to continuing experimentation with a number of alternatives. Kote, on the other hand, was easily divided into the two galvanizing lines of seven or eight people each.

### Employee Selection

The new operations faced an employee challenge: They needed experienced people, yet did not want to carry over adversarial traditions into the new setting. The rural area of the plant location could not supply the experiential requirement. Even the craft people involved in building the plant had to be imported. Most of them came from Illinois and Ohio because the local construction industry could not supply the requisite skills. Those with steel experience, primarily for maintenance functions, were selected from volunteers out of the Indiana Harbor works. More of them were interested in moving and starting over in a new plant than expected. Whereas applicants were expected primarily from the cold mill, they responded from the blast furnace, coke oven, open hearth, and other departments as well. Inland Steel had an elaborate bidding system designed to select on the basis of experience and skill, but Tek and Kote wanted to select by cooperativeness as well. As human resource staff put it, "We didn't want Rambo types, even if they were super-qualified and could run the world by themselves." Initially, most of the maintenance personnel were employees transferred from Indiana Harbor, while most of the operators selected locally and trained from among people had never been in a steel plant. Indiana Job Service was invited in with a time-consuming and costly testing procedure for the prospective bargaining unit people. Every one of the 3,000 applicants, whether bargaining unit or management, went through an assessment center screening process. The survivors were then put through a role-playing exercise with three company observers choosing among the survivors, by concensus. Flexibility potential was the ruling criterion. The same process has been followed in subsequent hiring situations.

### Training

Training at both enterprises has been a saturation process including a skills matrix and a pay-for-knowledge system. The Japanese influence is credited for the philosophy that training is a way of life and a daily part of every job. Yet, the Japanese assigned to the new facilities agreed that the training there went beyond anything they experienced at home. As a new undertaking, there was no opportunity for the new employees to learn step by step on the job. The initial crew for I/N Tek was selected one year in advance of opening and spent thirteen weeks in Japan at the counterpart Hirohata mill for training and familiarization. The Japanese involved in planning, construction, and start-up brought with them training plans as well.

In addition to the necessary technical training, the initial employees, along with everyone since, were intensively trained in basic mathematics, college chemistry, and computer programming. Regardless of background, everyone started at the bottom and had to be certified skill by skill in order to move up. This philosophy is built into the pay-by-knowledge compensation structure. Inland Steel veterans reported that they received more training in a year at Tek and Kote than in all of their previous working lives.

In order to get the operation up and running, technical knowledge had to take priority; but social skills such as teamwork, team building, communications skills, and so forth received even greater emphasis. On a continuing basis, teams are scheduled to be away from the line at least one day per month for some team-building activity. Despite intensity, there is a waiting list to participate in training with each new course offered being filled to capacity. All training is on company time and much of it is on an overtime basis. In addition, employees participate voluntarily on various committees such as safety, wellness, crew mix, and social activity planning. Whether that enthusiasm will prevail over the long run remains to be seen.

Reflecting on the role of the Japanese partner in the intensification on training, Controller Dennis Freeland comments:

> Nippon has led us to make sure that we are providing the training for technical areas. From my observations, however, we are way beyond what Nippon even considers training for their people in some of the other areas. I don't even know if social skills is part of their training matrix. . . . Obviously, our training here is far beyond anything at the Harbor Works.

He further adds:

> The Nippon people were great in the transition and the startup but they are more hierarchical and traditional, which sometimes conflicts with a commitment to training and employee involvement. They sometimes have had a hard time accepting the extent to which our joint ventures allow teams to make and carry out decisions.
>
> On the other hand, Americans are often guilty of too much of an independent spirit. Everybody has to do it their own way. It is pretty difficult to get all of our people to agree on what is the best methodology. Training people who are independent thinkers into a mindset that

there is value in concensus and team concepts is critical. . . . In many cases that drives the Japanese into looking at things a little differently too. It is training yourself to be creative but also more structured and rigorous in what you do. In the long run, you will get better quality and better working conditions and always be looking for constant improvement.

As a result, he exults, "This place just continues to get better and better, exceeding operations throughout the facility, and training has become a way of life consistent with our vision."

## Compensation

Pay for the operations and maintenance people combines hourly rates and bonus participation. As of 1994, wages for unskilled material handlers ranged from $8.50 to $10.60 an hour. Material handlers could bid on the basis of seniority to enter the operator ranks. Six operator levels ranged from $12.50 to $15.40, based on pay for knowledge rather than assignment or seniority. Assignments within teams are made by the teams on the basis of knowledge, skill, and minimization of disruption. Maintenance people—mechanics, electricians, and instrument service—had five levels with the same pay range. Quarterly bonuses, based on output and profitability, are shared equally by all which, at the time, were $4,000 a quarter, making the bonus component a higher proportion of the lower categories of pay. Thus, people with base pay ranging from $17,000 to $32,000, leaving out overtime, would each be adding $16,000 to their annual earnings, which became a source of envy in rural Indiana.

## Labor Relations

Whether the Tek and Kote employees would be represented by the USWA was never in question; it was required by Inland Steel's agreement. However, the new plants had a very different relationship. The first local president and vice president were both former Indiana Harbor employees but had never held union office there. The current president was a local resident who had a small business management background and had never been in a steel mill before. A separate agreement had been negotiated with the joint participation of local and international union officials with substantial departures from the Inland Steel agreement.

For contract administration, a Joint Advisory Council (JAC) consisting of, on the management side, the president, the two

Japanese vice presidents, the two managers of operations of the two facilities, and the human resource manager. This group met monthly with the president and vice president along with one elected representative of the membership in each plant in order to discuss major human resource and labor relations issues and review company plans. The local leadership and elected plant representatives also met twice a month with the maintenance and operations managers to discuss and resolve immediate issues. Then they carried their recommendations to the JAC for approval if necessary.

The compensation structure, bound to knowledge levels rather than job descriptions and classification, required radical new thinking. The nearest approach to job description was occasioned not by collective bargaining, but by the requirements of the Americans with Disabilities Act to determine reasonable accommodation for hiring of the disabled. The elimination of seniority as promotion criterion, except at the transition from materials handler to operator, also occasioned the rethinking of steelworker traditions. Management would like to have made greater use of temporary employees but acquiesced to union opposition. Contracting out was allowed; however, temporary employment was not. At this point, the willingness of the employees to work overtime had reduced the need. Given the pay and benefit levels in comparison to those of the community, turnover had been low and the employees were oriented to lifelong employment. Therefore, amicable labor relations might have been expected of them, but international officials had also shown a cooperative spirit in making the adaptations neccesary to the very different conditions and expectations.

## Cultural Conflict

Perhaps because there was no preexisting plant culture, the Inland–Nippon joint venture seemed to have avoided the cultural conflicts that other international joint ventures had experienced. At the beginning, the American participants had already spent time in Japan becoming familiarized with customs as well as technology and procedure. Then, both Japanese and Americans gathered in an open cornfield to create a plant. The only building on the property was a small farmhouse. Trailers were moved on-site for housing. The farmhouse was the office with the president in one bedroom, the vice president in another, and the controller in the third. The director of planning was placed in the dining room, with the living room serving as the conference room. This arrangement made for constant interaction and learning together. Participants from both nationalities still involved were able to reflect on the learning experience. Cultural differences were evident, and still are, but under the circumstances they did not rise to the level of conflict. The Americans tended to be more intuitive and inclined to trial and error, perhaps because of a tradition of dealing with abundance. The Japanese were more cautious, wanting their decisions to be driven by information, data, measurement, and small-scale experiment; perhaps this was a product of a climate of scarcity over the century. Americans tended to move ahead with confidence to install an innovation, while the Japanese wanted to try new procedures in simulation to be sure that they worked before installing them. Whereas the Japanese tended to follow prescribed rules, the Americans preferred to follow their own preferences. Japanese executives accustomed to obedience had to learn to explain reasons and accept argument. They now admit that often their people accept direction without argument but then pursue their given task without enthusiasm because they do not agree with the decision. American subordinates often require more coaching, counseling, and explaining; however, once they agree to a course of action, they perform more vigorously. Americans had to learn that "yes" from a Japanese employee did not necessarily mean "I agree." Often it meant "I hear you."

With operations underway, the Japanese were more conscious of housekeeping. The American controller remembers the Japanese vice president going around picking up cigarette butts to set an example for good housekeeping. The Japanese were more safety conscious; as a result, at least one full week per year was dedicated to safety training which Americans believed did not merit a day at any other typical American mill. The training emphasis also began with the Japanese but now, both believe, supersedes that. Their long-range planning emphasis has tended to lengthen the American's time horizons. On the other hand, the Americans who spent time at the Japanese counterpart mill noted that many tasks and decisions concluded in the central offices in Japan occurred on the shop floor in the United States. Also, young people were able to move ahead more rapidly in the United States.

Though the Japanese complement had typically been 20 to 25 out of 600 at the two mills, the Americans agreed that their partner's influence had penetrated deeply. Nevertheless, the American influence necessarily predominated on the basis of size and location. Both Japanese and Americans agreed that the environment at Tek and Kote was unlike any steel mill in either Japan or the United States, which was the product of the Indiana cornfield location.

### Markets

Though serving Japanese transplants was a major motivation for Nippon's participation in the I/N Kote joint ventures, over half of the output of the joint venture was sold to traditional American purchasers, primarily to U.S. automobile manufacturers. In fact, like parents setting up their children in business, most of the customer base of the two new entities came at the expense of the joint venture partners. Nippon turned over all of what it had been exporting which included a higher volume to Chrysler, for instance, than Inland had enjoyed. Inland Steel, on the other hand, had been the major supplier to Honda's Marysvale, Ohio, plant and now, channeled through I/N Kote, all the latter was capable of delivering. The same was true for all other customers. The new highly efficient and high product quality entities became the supplier of choice for both Nippon and Inland Steel.

As a result, it is difficult to tell how profitable the joint venture has been. After the initial start-up learning curve, both of the new firms showed a profit each year. But since Indiana Harbor operated below capacity and experienced losses throughout the 1990–1993 period before returning to profitability in 1994, some part of the early Tek and Kote profits may have been, in that sense, a gift from Inland.

To the extent which was possible, the new suppliers attempted to serve their customers with "just-in-time" delivery. Though this principle was learned from the Japanese originally, it is now standard practice in the American industry. It is a matter of cost, saving investment in inventory, as well as handling and storage costs which means something different in the case of steel than it does, for instance, in automobile parts.

Steel is an inherently lot-size manufacturing process. Though it is more flexible than it was in the past, it is still necessary to produce a substantial volume of one size and specification before making the major transitions required to begin a new lot of a different specification. It would require a continuous demand for an identical product to a single customer in order to feed directly off the producer's line to the customer's line. That, of course, is what exists and occurs between Indiana Harbor and I/N Tek. This phenomenon probably exists nowhere else between a steel company and an automobile company.

Nevertheless, it is now rare for an automobile manufacturer to have on hand more than a two-day supply of steel. Earlier, the automobile firms kept substantial inventories or depended upon warehousing middlemen so as not to be vulnerable to supplier interruptions; now they expect delivery guarantees. This shifts pressure to the supplier to maintain the inventory at the risk of discovering a defect which requires quarantine of the inventory until the flawed product is sorted out and discarded, or having a customer make a design change and shift away from the inventory on hand. In other words, warehousing at the producer's location can serve as an important guarantee of product quality. The automated storage and retrieval systems installed at Tek and Kote have made a major contribution to the "just-in-time" process, but substantial storage somewhere in the system is still essential. It is the intent of the two new suppliers to so plan, schedule, and earn customer trust in the years ahead to ship direct from theirs to their suppliers lines without warehousing costs to either.

### Product Quality

Product quality was, of course, first priority from the beginning for Tek and Kote. A TQM system was introduced and rigidly followed including prescription for the performance of every task down to how to type a letter or hold a wrench. TQM also required a radical change in maintenance procedures from those standard at Indiana Harbor. Those to be responsible for maintenance were sent to Japan to study Nippon's procedures of proactive, condition-based maintenance and came up with what at Tek and Kote is called the inspection method of maintenance. Rather than wait for breakdowns, which was the Indiana Harbor style, equipment would be constantly monitored and supported by careful record-keeping. As part of their regular inspections, maintenance personnel would, on a regular basis, interview the operating people at their work-sites to identify any impending problems. A consultant engineering firm with sophisticated equipment would come in each month to monitor each section of the lines on a rotating basis. Whether problems are identified or not, critical sections of each line were taken down and refurbished on a regular basis. Nine turns every month are dedicated to maintenance with all production ceasing. This is accompanied by a twenty-one-day major maintenance shutdown which occurs every two years and is written into the annual plans. One significant result has been a substantial reduction in the inventory of spare parts necessary to keep on hand in case of a breakdown.

### Technology

A preventive maintenance emphasis is essential to a continuous process where an interruption in one area can shut down the entire operation. Certain extensions and bypasses were built into the lines in order to allow for some delay in some components, but only for moments. For instance, portions of the line can be lengthened to allow for pulling out a roll and replacing it without shutdown, but this must be accomplished in a few minutes. The carousel, which was installed primarily to allow for diversion of product from Tek to Kote, would not only facilitate inspection, but also allow a shutdown in the annealing or skinpass operation without shutting down the pickling and tandem mill processing.

A continuous process contributes to quality. I/N Tek engineers point out that traditional tandem mills create blemishes and slivers by varying in speeds from 5,000 feet a minute to zero and also by threading and tailing out. The continuous process technology avoids this and also involves less waste as well as a longer life for equipment. The continuous process also offers great savings in absence of the inventory which otherwise would have to be maintained at various points on the line. However, the gains in efficiency and cost are bought at the price of vigilance and planning. As one manager said, "You are betting lower cost if you keep it running against the high cost of a shutdown if you foul up."

The Americans credit the Japanese for not only the development of the system but also for the guidance in installing it as well as the training in using it. As one manager at Tek and Kote put it:

> A few places have tried to install Japanese technology without Japanese help and it hasn't worked. It was more than buying the technology; by getting a partner instead of a supplier, you get some experience in its use that the machinery itself would not have provided. They sent some of their best people with years of experience to install the equipment,

work out the bugs and get it up and going. Some of them were here from six to nine months working hard to make our facility even better than their counterpart mill.

Both nationalities are convinced that substantial improvements have been made at Tek in comparison to Hirohata, but it is not all one-sided. Both also agree that while the steelmaking equipment was Nippon's contribution, the business system technology was Inland Steel's, going far beyond anything available in Japan.

### Assessment

The success of I/N Tek and I/N Kote is unassailable. It remains only to ask why. Location is a key element. Even though space was available at Indiana Harbor, both partners were committed to a new environment, both in terms of culture and business relationships. From an economic perspective, the new location was strategic to all of the principle customers in the automotive and appliance industries: automobiles in Detroit and Ohio; office furniture in Grand Rapids; appliances in Ohio; and Japanese transplants in Ohio, Southern Indiana, Illinois, and Tennessee. Transportation is also an advantage since none of the customers were located on water. These included a good road system, easy access to the toll road, and the ability to ship heavy loads by truck and rail. Utility costs are substantially lessened. Sitting on a water acquifer is another advantage. Air pollution is not an issue. The local polititical and business community are enthusiastic supporters.

The major location advantage is the workforce, which is free of cultural baggage. Though the nucleus of the workforce was from Indiana Harbor, it was carefully selected from those anxious for a new experience. The rest were rural and small town Indiana residents with no steel and little heavy industry background. Pay, benefits, and job security were beyond their expectations. Team concepts, employee empowerment, and other factors were as new to the recruits as to the Harbor transplants; however, the former had no concerted prior expectations. The steelworkers' union seems to have been amenable to the changing relationship but could not have recreated the Indiana Harbor labor relations environment in the Indiana cornfield, even if that had been its desire. The new environment was neither Inland Steel nor Nippon. There was heritage from both, but what emerged was a new creation.

This creation could not have occurred without a sacrifice from the parents. As noted earlier, Inland Steel did not itself make a profit during the early life of I/N Tek and I/N Kote. The two new entities operated at full capacity throughout and, with the exception of the start-up period, had been profitable. To what extent this was at Inland Steel's expense is difficult to say. Inland Steel had both cold mill and galvanizing capacity at Indiana Harbor, but they were not capable of the Tek and Kote quality. Therefore, some part of the sales of the new facilities are undoubtedly at Indiana Harbor's expense. In addition, the Tek and Kote operations are backed by the entire Inland Steel sales force. Besides, Inland Steel spent millions improving the Harbor facilities in order to improve the quality of the hot band substrate, which was supplied for processing by the new facilities.

Nippon also supplied cash and credit that would have not otherwise been available to finance the new ventures. It might have been possible for Nippon to buy the land, import the equipment, hire an American workforce, and, attracting orders from the transplants, go into business on its own. However, the American partners undoubtedly had an advantage in all but the transplant contacts, as well as providing a ready-made source of supply. Nippon's technology and its philosophical commitments to long-range planning, safety, and other attributes already enumerated were essential to the success of the new ventures. However, both the Japanese and Americans involved recognized that Tek and Kote had progressed beyond those beginnings. At the same time, the business and information systems of Tek and Kote were independent developments which were not available from Japan, and the team implementation and employee empowerment were home-grown developments. All of this amounts to a synergistic development which could not have emerged without the joint venture and which had also demonstrated its capability to progress beyond its auspicious beginnings.

## NOTES

1. William T. Hogan, *Capital Investment in Steel: A World Plan for the 1990s* (Lexington, Mass.: Lexington Books, 1992), 145; Martin Kenney and Richard Florida, *Beyond Mass Production: The Japanese System and Its Transfer to the U.S.* (New York: Oxford University Press, 1993), 156; "California Steel tie gives Kawasaki slab outlet," *American Metal Market*, October 19, 1984, pp. 1–8; along with site visit by co-authors Mangum and Kim, interviews with company principals, and company provided data.

2. William T. Hogan, *Minimills and Integrated Mills: A Comparison of Steelmaking in the United States* (Lexington, Mass.: Lexington Books, 1987), 29–30.

3. Dennis J. Fuga, "Takehiko Haga blending three cultures into success at California Steel," *The CSI: A Newsletter for Our Employees* 11(2): 1.

4. Scott Robertson, "Blending cultures at Pittsburg, California," *New Steel* (July 1994): 16–20; Hogan, *Capital Investment in Steel*, p. 207.

5. From site visits by co-authors Mangum and Kim, interviews with company and union officials, and company provided data. Sae-Young Kim offers special thanks to Executive Vice President Hak-Bong Ko, Director Won-Il Chun, and Manager Diana M. Mason of UPI for infinite patience in answering innumerable questions and providing useful materials and data. Kim also expresses gratitude to Senior Managing Director Jung-Woong Chang from the Korean Iron and Steel Association, General Manager Soung-Sik Cho of the corporate planning team of POSCO, and Chief Representative Kyung-Hwa Kim of the New York office of POSCO for their useful ideas, materials, and encouragement.

6. Allan Pre, "Outside(rs) in and Inside(rs) out: South Korean capital encounters organized labor in a California industrial suburb," in *Reworking Modernity*, ed. Allan Pre and Michael John Watts (New Brunswick, N.J.: Rutgers University Press, 1992), 155–164.

7. Robertson, "Blending cultures," pp. 18–19.

8. From site visits by co-authors Mangum and Kim, interviews with company and union officials, company provided data, and published sources.

9. "History of Inland Steel," in *The American Steel Industry 1850–1970*, ed. Kenneth Warren (New York: Oxford University Press, 1973), 313.

10. Seiichiro Yonekura, *The Japanese Iron and Steel Industry, 1850–1990* (New York: St. Martin's Press, 1994), 234–237, 256–258.

CHAPTER

7

# Lessons from Steel Industry Joint Ventures

The six joint ventures which were reviewed in Chapters 5 and 6 during 1994 and 1995 were part of a much different steel industry environment than the late 1980s environment which motivated the emergence of the joint ventures. The 1995 industry condition will be reviewed in relation to the continuing metamorphis of these firms before reflecting upon the apparent lessons from the steel industry joint ventures.

## THE STEEL INDUSTRY IN 1995

The year 1986 proved to be a pivotal one in the history of the American steel industry. Though it could not be known at the time, world steel production, U.S. shipments, capacity utilization, and net income all struck bottom (Table 7.1).

### Technological Response

Because of being deeply threatened, the firms in the industry took drastic steps to survive. At least fifteen major obsolete mills were closed during the late 1980s. Other plants were partially closed, which reduced overall industry capacity by 23.3 million tons between 1984 and 1988. This trend had much to do with a

**Table 7.1**
**Production, Capacity, Utilization, and Income, 1984–1993 (thousand tons, percentage, and millions of dollars)**

| Year | World Production | U.S. Production | U.S. Capacity | U.S. Capacity Utilization | Net Income |
|---|---|---|---|---|---|
| 1994 | 800,056 | 100,579 | 108,100 | 93.0 | n.a. |
| 1993 | 804,869 | 97,877 | 109,900 | 89.1 | $1,858 |
| 1992 | 810,229 | 92,949 | 113,100 | 82.2 | ($4,068) |
| 1991 | 813,238 | 87,896 | 117,600 | 74.7 | ($2,042) |
| 1990 | 846,600 | 98,906 | 116,700 | 84.7 | $54 |
| 1989 | 865,348 | 97,943 | 115,900 | 84.5 | $1,596 |
| 1988 | 858,932 | 99,924 | 112,000 | 89.2 | ($567) |
| 1987 | 811,367 | 89,151 | 112,200 | 79.5 | $1,077 |
| 1986 | 786,625 | 81,606 | 127,000 | 63.8 | ($4,149) |
| 1985 | 792,945 | 88,259 | 133,600 | 66.1 | ($1,834) |
| 1984 | 782,590 | 92,528 | 135,300 | 68.4 | ($31) |

*Source:* American Iron and Steel Institute, *Annual Statistical Report*, various years.
*Note:* Figure for 1994 world production is projected.

rise in capacity utilization. In addition to the joint ventures, several plants were sold to employees or to others from outside the industry. After lagging behind in the adoption of new technology, the U.S. industry began to make up for lost time as recounted in Chapter 4. These results are shown in Table 7.2.

## Employment Costs

The most obvious industry event of 1986 was the altered collective bargaining environment. With so many changes impacting many firms very differently, the industry's common front was shattered, but not to the advantage of the steelworkers' union. The year 1983 proved to be the last year of industrywide bargaining. In 1985, in anticipation of the 1986 negotiations, the Basic Steel Industry Committee was formally dissolved, and the union was informed that 1986 would be "every man for himself." It was company by company bargaining but the pattern was constructed of union concessions rather than offered benefits. The

**Table 7.2**
**Percentage of U.S. Raw Steel Production by Furnace and Cast Type**

| Year | Open Hearth | Basic Oxygen | Electric | Ingots | Continuous Casting |
|---|---|---|---|---|---|
| 1994 | 0.0 | 60.7 | 39.3 | 10.5 | 89.5 |
| 1993 | 0.0 | 60.6 | 39.4 | 14.3 | 85.7 |
| 1992 | 0.0 | 62.0 | 38.0 | 20.7 | 79.3 |
| 1991 | 1.6 | 60.0 | 38.4 | 24.1 | 75.8 |
| 1990 | 3.5 | 59.1 | 37.4 | 32.5 | 67.4 |
| 1989 | 4.5 | 59.6 | 35.9 | 35.1 | 64.8 |
| 1988 | 5.1 | 58.0 | 37.2 | 38.6 | 61.3 |
| 1987 | 3.0 | 58.9 | 38.1 | 40.1 | 59.8 |
| 1986 | 4.1 | 58.7 | 37.2 | 44.7 | 55.2 |
| 1985 | 7.3 | 58.8 | 33.9 | 55.5 | 44.4 |
| 1984 | 9.0 | 57.1 | 33.9 | 60.3 | 39.6 |

*Source:* American Iron and Steel Institute, *Annual Statistical Report*, various years.

union had begun piecemeal concessions to keep weak firms alive after 1983 and was now willing to spread those wage and work-rule concessions throughout the industry in exchange for its higher priorities which were job preservation, pension protection, limitations on contracting out, and successorship clauses to guarantee continued unionization of plants that were purchased by new owners.

Peaceful concession settlements were reached everywhere. However, USX pushed the USWA into the first labor stoppage since 1959, a stoppage that continued for six months, which was even longer than that year's other record of 116 days. Stoppage is said advisedly because at many plants a legal issue arose over whether a strike or a lockout was the appropriate cause of action. Several U.S. Steel plants were not reopened after the settlement was reached. The 1989 negotiations continued the concession process with the union taking a "leveling-the-playing-field" approach to keep the weaker companies alive and the stronger companies complaining about the rewarding of poor management.

As a composite result of technological upgrading, the closing of obsolete plants, work rule flexibility, and better management, output per employee hour rose by 40 percent between 1986 and 1993 after already having risen 38 percent between 1980 and 1986 (Table 7.3). Steel industry employment was cut in half within twenty years. Hourly employment costs stabilized while labor costs per ton of steel, aided by productivity improvements, plunged to 1990 and then fluctuated according to capacity utilization levels (Table 7.4).

## International Competition

There were dual causes for the rise of U.S. steel production after 1986. First, the economy was generally on an expansion track from the 1981 recession and was further sparked by rising government deficits in prosperous times, continuing until a relatively mild recession during 1991–1992. The import picture also improved, both from restrained U.S. costs and because the

**Table 7.3**
**Productivity Index (1987 = 100)**

| Year | Employment | Employee hours | Output | Output per employee hour |
|---|---|---|---|---|
| 1970 | 233.7 | 215.6 | 124.2 | 57.8 |
| 1975 | 204.3 | 186.2 | 114.1 | 61.3 |
| 1980 | 190.8 | 173.3 | 117.0 | 67.5 |
| 1985 | 114.2 | 106.9 | 97.7 | 91.4 |
| 1987 | 100 | 100 | 100 | 100 |
| 1988 | 103.8 | 104.2 | 114.9 | 110.3 |
| 1989 | 104.0 | 103.4 | 110.8 | 107.2 |
| 1990 | 102.9 | 102.3 | 112.9 | 110.4 |
| 1991 | 97.9 | 95.9 | 101.9 | 106.3 |
| 1992 | 93.3 | 92.6 | 107.5 | 116.1 |
| 1993 | 89.0 | 89.5 | 116.8 | 130.5 |
| Average annual change | | | | |
| 1970-86 | -8.4 | -7.2 | -1.5 | 2.7 |
| 1987-93 | -1.8 | -1.8 | 2.8 | 5.1 |

*Source:* Bureau of Labor Statistics Bulletin 2461, May 1995.

**Table 7.4**
**Employment and Labor Costs**

| Year | Average Employees | Employment Cost per Hour | Employment Cost per Ton |
|---|---|---|---|
| 1994 | 125,618 | $33.71 | na |
| 1993 | 127,192 | $32.30 | $88.54 |
| 1992 | 139,664 | $29.78 | $90.00 |
| 1991 | 146,140 | $28.16 | $95.33 |
| 1990 | 163,963 | $26.34 | $83.64 |
| 1989 | 168,853 | $25.35 | $84.57 |
| 1988 | 168,898 | $25.30 | $88.54 |
| 1987 | 163,338 | $24.23 | $87.07 |
| 1986 | 174,783 | $23.57 | $92.44 |
| 1985 | 208,168 | $23.26 | $99.96 |
| 1984 | 236,002 | $21.82 | $100.88 |

*Source:* American Iron and Steel Institute, *Annual Statisticsal Report*, various years.

dollar, which had been excessively strong vis-à-vis the yen, and the mark began to weaken, encouraging exports and discouraging imports. Lowered U.S. capacity raised utilization rates and restrained production costs; however, elimination of excess capacity meant either shortages or accentuated imports in years of high demand such as 1994 (Table 7.5).

Notably, it was the newcomers on the world steel scene who were filling in at the margin (Table 7.6). U.S. hourly compensation costs became equal with those of Germany and substantially less than those of Japan. Further, Korean labor costs remained far lower (Table 7.7) because the Koreans had no excess capacity to export. U.S. employment costs per ton of steel were now below the costs of its major competitors, which were $150 per metric ton for the United States compared to $162 for the Europeans and $195 for the Japanese. However, in 1995, the figure for the Koreans was $70.[1] Except for periods of short supply, imports were now a fairly stable proportion of American steel consumption.

**Table 7.5**
**Sources of Steel Supply (thousands of net tons and percentage)**

| Year | U.S. Shipments | Exports | Imports | Imports as Percentage | Japan | EU | Canada | Other |
|---|---|---|---|---|---|---|---|---|
| 1994 | 95,084 | 3,826 | 30,066 | 24.8 | 3.0 | 7.7 | 3.8 | 10.4 |
| 1993 | 89,022 | 3,968 | 19,501 | 18.7 | 1.7 | 6.5 | 4.6 | 5.9 |
| 1992 | 82,241 | 4,288 | 17,501 | 18.0 | 2.8 | 4.8 | 4.5 | 5.9 |
| 1991 | 78,846 | 6,346 | 15,845 | 17.9 | 3.2 | 5.3 | 3.3 | 6.1 |
| 1990 | 84,981 | 4,303 | 17,169 | 17.5 | 3.2 | 5.6 | 3.0 | 5.7 |
| 1989 | 84,100 | 4,578 | 17,321 | 17.9 | 3.8 | 5.8 | 3.1 | 5.2 |
| 1988 | 83,840 | 2,069 | 20,891 | 20.3 | 4.2 | 6.1 | 3.1 | 6.9 |
| 1987 | 76,654 | 1,129 | 20,414 | 21.3 | 4.5 | 6.0 | 3.8 | 7.0 |
| 1986 | 70,263 | 929 | 20,692 | 23.0 | 4.9 | 7.3 | 3.6 | 7.2 |
| 1985 | 73,043 | 932 | 24,256 | 25.2 | 6.2 | 7.3 | 3.0 | 8.8 |
| 1984 | 73,739 | 980 | 26,163 | 26.4 | 6.7 | 6.4 | 3.2 | 10.1 |

*Source:* American Iron and Steel Institute, *Annual Statistical Report,* various years.

### Mini-Mill Competition

The intense competition in the United States was now internal between the integrated mills and the mini-mills. The advantage of the mini-mills was based on a combination of labor cost advantage, which was $38 per ton in 1995, low capital investment costs, and expansion beyond low-end products such as bar, billets, rail, and rods into thin-slab casting for some flat-rolled products. However, their market share was limited to about one-fifth because of scrap availability (high scrap prices which brought their costs near to the best of integrated mills in times of high demand), and the inability to produce an unblemished, high-end, flat product. Those limitations are subject to change over time, however, as is evident when observing the experiments of Nucor Steel with iron carbide production and other experiments with direct reduction of iron, trending away from sole dependence on scrap.[2]

### New Directions in Collective Bargaining

Recognizing industry survival as essential to its own survival as well as the economic well-being of all of those members whose jobs could be saved along with all of those retirees dependent upon the pensions and health benefits owed them, the USWA proposed, in 1993, an approach to collective bargaining which had no U.S. precedent. Union negotiators returned to the bargaining table in 1993 with a totally new approach which they described as a "New Directions Bargaining Program." The union offered long-term contracts, further relaxation of restrictive work rules, and cooperation in further attrition reduction of steel company workforces. They wanted, in return, increased job security for their membership, adequate funding arrangements for the "legacy costs" of pensions and health-care benefits for the retired, and a stronger voice in the management of the firms. Couched in the language of "twelve points of light," these bargaining goals could be summarized as the following:[3]

- Long-term agreements (six years in most cases)
- Enhanced job security, including guarantees against layoffs except under disaster conditions as well as strengthened prohibitions against subcontracting and excessive overtime
- Reduction of health-care costs through the establishment of managed-care networks
- Union and worker empowerment ranging from international union representation on boards of directors to joint shop-floor committees, all without dilution of union bargaining rights
- Company neutrality in union organizing campaigns along with union recognition at new facilities based on authorization card checks
- Third-party resolution of subcontracting disputes

The remarkable fact was that all of the major basic steel companies agreed after some initial hesitation on the part of USX and without even a strike threat. Only time will tell how the radical new departure will work out in practice and what the results will be; however, the promise is a new dawn in a historically adversarial labor-management relationship.

### The Outlook

Turning to the previously dismal picture of profit and loss, in 1987, 1989, and 1993, a corner appeared to have been turned

**Table 7.6**
**World Steel Production, 1984–1993 (thousands of metric tons)**

| Location | 1984 | 1986 | 1988 | 1990 | 1991 | 1992 | 1993 |
|---|---|---|---|---|---|---|---|
| World | 710,267 | 713,998 | 780,092 | 770,141 | 736,150 | 723,442 | 730,644 |
| Western Europe | 156,994 | 149,714 | 164,307 | 162,825 | 161,811 | 157,054 | 158,096 |
| Eastern Europe | 60,203 | 61,321 | 61,372 | 49,123 | 33,161 | 29,459 | 29,788 |
| Former USSR | 154,238 | 160,537 | 163,137 | 154,414 | 132,839 | 117,977 | 97,849 |
| North America | 106,813 | 96,219 | 114,105 | 111,447 | 101,391 | 107,486 | 113,044 |
| Mexico | 7,560 | 7,225 | 7,779 | 8,734 | 9,964 | 8,459 | 9,199 |
| South America | 25,436 | 29,779 | 34,093 | 29,049 | 30,878 | 33,051 | 33,722 |
| Brazil | 18,386 | 21,233 | 24,657 | 20,567 | 22,617 | 23,934 | 25,170 |
| Africa | 10,318 | 12,520 | 13,049 | 13,359 | 15,175 | 14,427 | 14,002 |
| Middle East | 2,373 | 2,680 | 3,321 | 4,064 | 4,779 | 5,589 | 6,874 |
| Iran | 854 | 838 | 978 | 1,425 | 2,203 | 2,937 | 3,672 |

| | | | | | | | |
|---|---|---|---|---|---|---|---|
| Saudi Arabia | 842 | 1,100 | 1,614 | 1,790 | 1,783 | 1,823 | 2,318 |
| Asia | 187,311 | 194,257 | 219,538 | 238,466 | 249,126 | 250,726 | 268,585 |
| Japan | 105,586 | 98,275 | 105,681 | 110,339 | 109,649 | 98,132 | 99,723 |
| China | 43,475 | 52,208 | 59,430 | 63,349 | 71,000 | 80,935 | 89,539 |
| South Korea | 13,034 | 14,555 | 19,118 | 23,125 | 26,001 | 28,055 | 33,029 |
| India | 10,549 | 12,197 | 14,309 | 14,963 | 17,100 | 18,117 | 18,155 |
| Taiwan | 4,934 | 5,545 | 8,288 | 9,748 | 10,973 | 10,705 | 11,970 |
| Oceania | 6,582 | 6,961 | 6,947 | 7,395 | 6,990 | 7,652 | 8,683 |

*Source:* Adapted from International Iron and Steel Institute, *Steel Statistical Yearbook 1994* (Brussels: International Iron and Steel Institute, 1995), pp. 23–24.

**Table 7.7**
**Hourly Compensation Costs for Production Workers (Index U.S. = 100)**

| Country | 1986 | 1990 | 1993 |
|---|---|---|---|
| United States | 100 | 100 | 100 |
| Canada | 70 | 87 | 78 |
| Japan | 74 | 95 | 112 |
| Korea | 10 | 26 | 32 |
| Germany | 70 | 102 | 100 |

*Source:* U.S. Bureau of Labor Statistics.

(Table 7.8). In each year except for 1993, the promise proved a disappointment. At least, the 1993 upturn continued through mid-1995. Although final data was not available before this book was printed, 1994 was a remarkably good year with the U.S. industry running flat out and imports zooming to meet demands which were not within the reach of the downsized American capacity. No industry suffering losses in half of the past ten years could yet declare its health to be good; however, all of its major ills had been addressed in that same decade. Its technology had been modernized, its productivity accelerated, its labor cost sharply cut, its international competition confronted and restrained, and its profits rejuvenated.

However, even in the best years, the rate of return in a mature industry such as steel cannot be impressive (see Table 7.8). In a cyclical industry, it is always difficult to separate temporary recovery from long-range trends. The mature industry must continually struggle to maintain its share of a relatively fixed steel consumption demand, fluctuating cyclically but not rising secularly. Nevertheless, there is good reason to think that the U.S. steel industry had at least fought its way back to viability. International joint ventures were no small part of that turnaround.

## JOINT VENTURE LESSONS

The six case studies in Chapters 5 and 6 provide some rather clear lessons which are supported, in less detail, by the information in Chapter 4. First, like the course of true love, none of the industrial marriages were without strain. Even though two eventually ended

**Table 7.8**
**Income Data (millions of dollars)**

| Year | Net Sales | Net Income | Percentage of Net Sales | Total Assets | Percentage of Total Assets |
|---|---|---|---|---|---|
| 1993 | $29,682 | $1,858 | 6.3 | $30,714 | 6.0 |
| 1992 | 26,855 | (4,068) | -15.1 | 28,803 | -14.1 |
| 1991 | 26,871 | (2,042) | -7.6 | 27,437 | -7.0 |
| 1990 | 30,635 | 54 | 0.2 | 28,257 | 0.2 |
| 1989 | 31,525 | 1,597 | 5.0 | 24,691 | 6.6 |
| 1988 | 32,466 | (567) | -1.7 | 24,187 | -2.3 |
| 1987 | 26,933 | 1,077 | 4.0 | 22,198 | 4.9 |
| 1986 | 24,875 | (4,150) | -16.7 | 20,957 | -19.8 |
| 1985 | 28,272 | (1,834) | -6.5 | 23,950 | -7.6 |
| 1984 | 30,005 | (31) | -0.1 | 26,187 | -0.1 |

*Source:* American Iron and Steel Institute, *Annual Statistical Report*, various years.

in divorce, the separations were amicable. None of these liaisons were unproductive and every one saved an industrial enterprise that was otherwise slated for death. Further, a major economic contributor and source of employment was saved for every community in which these steel facilities were located. With the exception of I/N Tek and I/N Kote, which were totally new greenfield creations, every plant did experience reduced employment, but the alternative would have been no employment at all.

There is no reason to think that there would otherwise have been an I/N Kote and I/N Tek in a cornfield in New Carlisle, Indiana, without that joint venture. Much of the finished products of the two Western mills would probably have been imported from the same two countries which provided the saving capital rather than by simply importing the semi-finished product. The steel products of the midwestern joint ventures would otherwise have been produced elsewhere in the midwestern United States, though perhaps not with the same quality or with the same efficiency. Steel prices would have undoubtedly been higher with the advantage going to competing producers, but disadvantage going to the consumers. Thus, the economic contribution may have been more local than national, but it was significant to both.

### Achieving Objectives

All of the participants achieved their objectives. For the American partners, the primary objective was financial investment to support modernization. None of the enterprises were attractive to either lending bankers or equity investors under the prevailing circumstances. All obtained the "patient capital" they sought, which were loans from those willing to take the risk and accept interest rewards along with slow amortization. However, this patience proved to have its limits; but as long as interest obligations were met and the continuance of interest payments with timely amortization was reasonably assured, that patience endured. However, as soon as either interest payments or capital repayment appeared doubtful, the international bankers became not only impatient, but fearful and "blew the whistle." This happened in two of the three integrated mill cases, and, in each case, the owners' refuge was the supposedly impatient stock market. Fortunately, the joint venture had brought the entities to a promising stage, and the Japanese bankers, who were plagued by a recession at home, bailed out just when steel profits, or the promise thereof, were again attractive to U.S. investors.

For one of the integrated, mill-owning U.S. firms, National Intergroup, Japanese money was a second-best solution. They wanted out of the industry altogether, and that was ultimately what they got. In the end, the long-term outlook for the steel mills may have been better than that of some of their preferred alternative investments, but at least NII was able to sell out of steel for funds in order to bail out other wavering subsidiaries. All the other American joint venturers ended up with soundly financed enterprises, and, in turn, they got much more in the bargain. All of them, including National Steel, were managed much better after than before the joint ventures and their aftermath.

The Japanese all wanted entry into the world's largest steel market without the political opposition generated by imports. They were also being pushed by their domestic automobile customers, who had already transplanted themselves into the world's largest automobile market, for the same reason. Steel was only part of a supplier complex which included automobile parts, tires, and even machinery production which were created under familiar and trusted management in the United States. The U.S. investment may have been part of a larger plan for self-sufficient regional industrial complexes on a global scale.[4] All would probably have preferred to supply the world from their home facilities through export to the extent possible, but could see the political handwriting on the wall for one-sided policy.

The transplants all accomplished their U.S. objective, though the advantage to their parent companies is not necessarily proven. However, NKK at National Steel and Kawasaki at AK Steel ended up with majority stock ownership without meaningful operating roles in American steel firms. That ownership status is a substantial departure from the intended two models which were either to buy into existing facilities and modernize them or to join in creating totally new facilities, but, in either case, to remain integrally involved in both policy and operations. The advantages of majority ownership when accompanied by nonmanagerial status may be debatable, but a useful outpost is probably still provided.

The objective of Kawasaki, along with its Brazilian partners at California Steel, was to obtain a finishing mill outlet for its Brazilian iron ore and slab. This was clearly accomplished. POSCO needed an outlet for steel capacity imposed upon it by the Korean government for political job creation objectives. That too was accomplished, though there is now sufficient domestic and Far Eastern demand to absorb the total production of POSCO's two domestic plants if it were not for the U.S. commitment.

The five Japanese firms, responding to the pressures from their Japanese auto transplant customers, met those requests, though NKK and Kawasaki are no longer directly involved in production and delivery. No one on either side has reason to regret their joint venturing decisions. The transplants will probably continue their familiar relationships, despite the reduced involvement of their steel counterparts, without a quiver.

### Challenges for Integrated Mills

The challenges to the integrated mills were clearly more intense than those to the finishing mills. All three were aging facilities which required large-scale modernization, upgrading, and expansion into coated products. All three had been deficient in meeting the rising quality demands of their customers. All had traditional workforces represented by unions immersed in adversarial bargaining. Finally, in all three cases, the labor relations stance of the management had long been as adversarial as that of the unions. Indeed, the stance of the unions had emerged as the mirror image of management's responses to employee needs.

The Kobe/U.S. Steel joint venture at Lorain, Ohio, was immediately successful for two reasons. First, it was profitable and engaged in the production of products wherein the competition was not intense—bar and tubular rather than flat-rolled steel. Its parent's inability to provide the funds needed for modernization was the result of corporation-wide weaknesses, not the Lorain outlook. Second, the Lorain plant management who were in place before the joint venture along with many of the American managers who came on board at the time of the joint venture were already committed to the employer–employee relations principals fostered by the Japanese–American partnership. Beginnings of participatory managment were already in place which meant that a conversion period was not necessary. In the National Steel and Armco cases, the joint ventures could not solve the basic problems, but they could, however, guarantee survival until more unilaterally American solutions could be mounted. In the National Steel case, that was the exit of a reluctant partner, and at AK Steel, it was the advent of tough-minded management. Ultimate survival under new American management at both entities remains to be tested; however, things appear to be promising.

### The Finishing Mills

The task at the three finishing mills was more like the task at the Lorain integrated mill. All three faced eager markets and employed willing workforces; however, the challenge of responding to that demand was primarily technological. The West Coast canning and appliance market for the U.S. Steel/POSCO joint venture and the West Coast construction market for CSI were a long distance from eastern and midwestern mills. The relevant competition was the steel industries of the nations with which the Americans, which in one case meant employees only and in the other meant owners and employees, were allying themselves. Both western facilities needed only funding, guidance, and the provision of technology familiar to their joint venture partners to meet the intensifying quality demands of their customers—demands which had, in part, been generated by the quality demonstrated by the Far Eastern competitors. I/N Tek and I/N Kote were totally new plants, bettering even the leading technology provided by their Japanese partner and were positioned to be leaders in an expanding market. They also had a new and unsullied workforce and were given priority access to both sales staff and customers of their Inland Steel parent.

CSI enjoyed the flexibility of non-union operation, endorsed by a workforce which continued to reject union organizing gambits. The other two finishing mills, while unionized, had, in one case, a workforce largely free of collective bargaining traditions, and, in the other case, gratitude for the continuance of employment that had looked doomed. Both were far enough away from other steel centers that the international union was not concerned with departures from eastern and midwestern union patterns. All of the joint ventures occurred in an era wherein the battered union was more amenable to compromise than had been its history. In fact, at the end of the period the union had vaulted into an unprecedented position of industry leadership in labor–management relations experimentation.

### Unanticipated Bonuses

The American partners primarily sought and received access to capital but were rewarded with much more than that. Every plant, whether integrated or finishing, was exposed to, and adopted a new commitment to quality. The first step was technological improvements. Only I/N Tek and I/N Kote had complete freedom technologically, but all the others were rewarded by major upgrading and modernization of existing facilities as well as the addition of new facilities. However, it was more than technology; there was intense planning, detailed specification of standards, strict inspection and careful quality control, careful equipment maintenance, good housekeeping, constant preaching, and a reward system which placed cash value on quality. All participant firms eventually won plaudits from customers for their quality improvements, and some even received added orders and exclusives for certain products. At this point, the new quality consciousness across American industry appeared to be deepseated and gave promise of a permanent change. That, in itself, is an international influence. Higher quality imports eventually forced U.S. producers to raise their standards. Higher quality standards for final products required higher quality raw materials and semi-finished products. Higher quality imported steels forced a higher quality domestic product.

Galvanized and other coated products are a prime example. The Japanese made greater use of coated steel in order to protect their automobiles from the excessive use of ice-melting salt in Japan. Accustomed to that supply, the Japanese auto transplants stated that they couldn't get it in the United States. The steel joint ventures emphasized this product and now no American auto manufacturer would be without it. If supply did not create its own demand, certainly competition did, and to the benefit of all.

A switch from the notion of "fix it if it breaks" to scheduled preventive maintenance was another universal gain for the joint venturers. Whether this practice will survive will be tested the next time that cash flow tightens and managers are looking for expenditures that can be temporarily delayed. All the joint venturers were also led by their Far Eastern partners into long-range planning. Its survival is probably more tenuous, especially for the two integrated mills which have now freed themselves from any substantial international role in their management.

Gains in human resource management are not as obvious. Lifetime employment gave way to cost-cutting needs at UPI, National Steel, and AK Steel while remaining untested at the other three joint ventures. AK Steel had experienced major manning reductions before the joint ventures and have faced no pressure to further downsize. The foreign partners resisted but finally succumbed to financial pressures. To the degree that their steel industries are concerned, neither the Japanese nor the Koreans have demonstrated strong commitments to employee participation in workplace rule making nor team approaches to production. Their method has been more concerned with the generation of loyalty through employment guarantees rather than the promotion of an employee voice. These participatory experiments were in place as much from American as foreign influence and commitment. The foreign partners had demonstrated no special gifts for labor relations in these long-unionized establishments. Only at UPI did the employee representatives show a preference for the foreign partner, but that probably reflected the animosity toward USX which was prepared to abandon them until the Korean saviors came to the rescue. Even there, labor-management negotiations remained the domain of the American partner. The system would have been booby-trapped for an uninitiated employer.

The union reaction is also worthy of note. The local unions were, of course, most concerned with the impact of the joint venture upon their local membership. They were favorably predisposed because they knew that without the joint ventures, their jobs would have disappeared. However, this did not eliminate militancy. Often, concessions were made to facilitate the joint venture negotiations. Even without particular local concessions, the national negotiations had involved concessions or limited gains. As soon as the joint ventures were well established, they

wanted rewards for their members. As someone stated, "We shared the pain; now we want to share the gain!" The local union officers also tended to be imbued with militancy. They were not usually the ones who were most amenable to labor–management cooperation. At the Lorain plant, for instance, local union leadership and work team leadership ended up in separate but noncompeting hands. Only at I/N Tek and I/N Kote, where there was no history to live down, were the new local union leadership and the new local management totally compatible.

At the international union level, the reactions were equally divided. All three integrated mills were by tradition, locality, and technology part of an industrywide system of long-standing. In the 1980s, the USWA learned the penalties of granting concessions to some and not all in a competitive situation.[5] Relations at National Steel and Armco were such that concessions, from what had once been industrywide bargaining but had now subsided to pattern bargaining, were not pursued on either side. USWA negotiators were convinced they could not allow U.S. Steel/Kobe to escape from the U.S. Steel pattern without repercussions throughout corporate negotiations. On the other hand, the international union was a positive force in working out new relationships in the finishing mills where no national pattern was threatened. The California mills were threatened because of distance and the Inland Steel/Nippon joint venture was threatened because of its uniqueness. Like any other institution, unions respond in accordance with their perception of impact on their raison d'être.

The great unknown in the steel labor relations picture is the impact of the startling 1993 and 1994 collective bargaining innovations. These were totally union initiatives. An industry once characterized by concerted price and wage leadership among its dominant firms, now appeared to be union-led into an unplumbed new era. The joint ventures appeared to have no evident role in bringing this trend about. The foreign influence was more likely the advent of a Canadian to USWA leadership. If the lack of joint venture input is true, it is paradoxical that partners from nations highly reputed for their superior employee involvement should be beneficiaries of such home-grown departures.

Trade relations were more clearly beneficiaries of the joint ventures. The voluntary restraint agreements, along with other U.S. protectionist policies since the 1960s, were accompanied by disinvestment and neglect of technological and organizational upgrading. American steel firms were using their facilities as

"cash cows" in order to support corporate diversification. The joint ventures, by establishing a viable new model of production organization, may have contributed at least modestly to some amount of reindustrialization. Not only capital but technology has been proven internationally mobile. The center of steel production technology had moved from Europe to the United States and then to Japan. Now such advanced Japanese technologies as electro-galvanizing and hot-dip sheet finishing are diffusing back to the the United States through international joint ventures and with notably less strife. Occasionally, one hears even the international joint venturers call for government protection, but certainly those voices are muted in comparison with the decibels heard during the 1980s.

The variety of relationships involved in these case studies provides insight into the requirements for successful international partnering. Though cultural differences were obvious, they produced no conflict where the joint venture was followed by a return to profitability, but did produce conflict where success was elusive. Obviously, there must be a shared need—the more intense the better. The partners must continue to need each other. Once the needs shift, the alliance becomes vulnerable. Objectives need not be identical, but they must be consistent. Above all, both partners must be fully committed. Joint venturing is no place for halfheartedness, as the National Steel experience demonstrates.

### Environmental Motives and Consequences

Only the CSI experience hints at what one might expect to be a major motivation for an IJV in steel production, namely, an upstream and a downstream mix that saves both transportation costs and environmental degradation. Long exploited U.S. iron ore supplies are declining in quality while, at the same time, there are ample supplies of rich ores in developing countries. Quality coking coal is still available in the United States, but that supply cannot last forever. Producing steel at the raw materials source, as in the Brazilian case, rather than shipping for thousands of miles the ten or so tons of raw materials required for every ton of steel slab has obvious advantages.

In addition, the environmental challenges of steelmaking primarily involve coke ovens and blast furnaces. Raw material supplies are generally found away from population centers in developing as well as developed countries. Poorer countries are

more likely to make a different calculation of the relative merits of income versus pristineness. Hence, one might expect a trend for steelmaking to occur in developing countries which are close to raw material supplies where the semi-finished product can then be sent to the consumer outlet for finishing.

Note that the CSI relationship fits both of these hypotheses, especially when one considers the finishing mill location in the environmentally sensitive Los Angeles Basin. Also note that the U.S. Steel/POSCO relationship does not fit. Korea is a small country without raw materials. The necessary raw materials have come from as far away as Australia and South America for processing in an environmentally sensitive environment before sending the output thousands of miles further to the United States. Were it not for the low quality of their iron ore, one would expect, over time, to see China become the raw material source and perhaps even the slab producer with finishing occurring in Korea and the United States. Japan and Taiwan also might eventually expect to become finishers rather than basic steelmakers.

On the other hand, direct reduction steelmaking, which is now largely experimental, may eventually bypass the coke oven and blast furnace stage with less negative environmental consequences. The 1994 joint venture between U.S. Steel and Nucor provides interesting dual signals.[6] Its ore is mined in Brazil and processed in Trinidad for steel production in the United States; undoubtedly, this has environmental gains for the United States. Further, its direct steelmaking bypasses the coke ovens and blast furnaces which would have been the worst of potential polluters to be diverted to less developed countries. It will be interesting to watch this industry drama unfold and see which script is followed. At any rate, international joint venturing does offer additional opportunities for rationalizing on many fronts.

## BACK ON THE INDUSTRIAL STAIRCASE

As was noted before the steel industry examples studied in Chapters 5 and 6, movement up the industrial staircase confronts advanced nations with the necessity of either speeding their ascendancy, or fighting it out with other venturers in order to retain position on each step of the industrial staircase. Joint venturing offers a peaceful and potentially productive vehicle for joint tenancy. Joint venturing also brings mutual accommodation to the classical economic adage that each nation should produce for trade those products in which it enjoys the greatest

comparative advantage. Rather than any notion of "all or nothing," that comparative advantage can be subdivided to production components or to elements of the production process. Joint venturing has also offered the steel industry a smoother and more accommodating approach to technology transfer and has reduced resistance to learning from different firms which, in turn, avoids hostility.

In no case did the foreign partner come on the scene prepared for unilateral success in the American environment. Neither did they pursue or achieve managerial dominance. In every case, the Japanese or Koreans brought money, better technology, and new commitments to advanced management principles. Further, the foreign staff came in very small numbers and individually stayed for relatively brief periods. They could participate in policy decisions and provide training and example, but they could not dominate the management or production processes. Even if they could have done so, it probably would not have been effective. The imported principles had to be translated into a U.S. environment; only the American partners could do that. In addition, whether by preselection or conversion, there was a clear tendency for the American management participants to be those who were already predisposed to adopting the very reforms their international partners were committed to bring. In most cases, the American managers—and in the employee relations arena, the union—appear to have improved on the techniques introduced or accentuated by their new partners. They also Americanized those policies and procedures.

As we noted earlier, the imagery of nations on an industrial staircase can be misleading; it is firms within nations which are the action point. Certainly, it was the individual steel firms who were looking for viability and pursuing their own self-interest who developed the IJV and discovered its contribution to mutual gain. It was the respective governments, which could only block or encourage, who could not innovate or instigate.

These joint venturing firms have done their respective nations proud. The foreign partners have learned helpful lessons for further globalization. One cannot argue that the U.S. steel industry would not have survived without globalization. It would certainly have been weaker had the joint venturing not occurred. Focusing on six joint ventures which involve only four American firms may camouflage the fact that every old-line American steel firm has become involved in these or other IJVs. Even from those where the partnership proved temporary, no sigh of regret has been heard.

In stark reality, the United States, in the lead position on the industrial staircase, has faced the choice between abandoning its steel industry to competitors or fighting a continuing battle. The United States made the latter choice and emerged leaner but stronger for the experience. The headline of a 1995 *Business Week* article illustrates the dramatic change from a decade earlier: "Can Anything Dent This Industry? The Big Three Carmakers and Emerging Foreign Demand Have U.S. Steel Mills on a Roll." This article states that the joint ventures were an effective tool in that survival and recovery.[7] As suggested in another 1995 *Business Week* article, the threatened components of the steel industry may have shifted away from the integrated sector.[8]

There seems to be no further need for more IJVs in the American steel industry's near future for defensive reasons; however, they may reoccur for environmental purposes. The fact that so many IJVs are occurring in the steel industry outside the United States at present demonstrates their usefulness for a variety of objectives. Will Japan share some of the same experiences in the near future as China surpasses it in steelmaking capacity? Further, will it not learn valuable lessons from the U.S. experiences which are to be relearned by others thereafter? It remains for us to extract contributions to joint venture theory and to explore what these lessons from the steel industry might offer to potential joint venturers across the economic spectrum.

## NOTES

1. Donald F. Barnett, "Strategic outlook—1995." Paper presented at Steel Survival Strategies Ten: The Changing Battleground, American Metal Market/World Steel Dynamics, New York, June 21, 1995. Table 3A is the source for U.S., European Union, and Japanese costs, whereas those for Korea are from POSCO Research Institute. Despite lower costs, Pohang Steel Company does not currently attempt to increase its exports to the United States because it is already operating at full capacity to meet the growing domestic demand on top of its existing international commitments.
2. Ibid., p. 13.
3. "Restructuring the United States steel industry through collective bargaining: USWA agreements designed to ensure the future of the integrated steel industry and long-term employment benefit protection for steelworkers." Paper presented by USWA representatives at the International Conference on Workplace Change: Human Resources and Rationalization in the Global Steel Industry, Toronto, June 5–6, 1995.

4. B. O. Huallachian, "The restructuring of the U.S. steel industry: Changes in location of production and employment," *Environment and Planning A* 25(1993): 1347.
5. John T. Hoerr, *And the Wolf Finally Came* (Pittsburgh: University of Pittsburgh Press, 1987), 475.
6. "The odd couple in steel: Rivals Nucor and U.S. Steel just may reshape how metal is made," *Business Week*, November 7, 1994, p. 106.
7. "Can anything dent this industry? The Big Three carmakers and emerging foreign demand have U.S. steel mills on a roll," *Business Week*, January 9, 1995, p. 68.
8. "The brutal brawl in steel. How many minimills is too many? The industry will soon learn," *Business Week*, March 13, 1995, p. 88.

# Chapter 8 Theoretical Implications

Chapter 7 drew lessons for the steel industry from the experience of six joint ventures. As such, Chapter 7 provides a practical analysis of these ventures and proposes some broader implications for managers who are considering international joint venturing. However, the chapter stays in its industry and does not attempt to draw conclusions beyond the evidence at hand. In this chapter, broader concepts are examined. The theoretical and conceptual bases proposed in Chapter 2 are revisited which involve forming and managing IJVs and interpreting the examples from the steel industry in light of this theoretical background.

By doing so, we intend to address two key issues. First, we are interested in understanding what the various organizational economic and sociology models can say to systematize and clarify just what happened to encourage the sudden development of production-focused IJVs in the U.S. steel industry. Second, we will investigate what this industry has to show about the relative value of these various general models in understanding a specific national industry context. By bringing general concepts into a specific situation, we hope to learn more about the situation and more about the concepts.

This analysis appears to divide logically into two major concerns. One is to examine the decision to form a joint venture. As stated in Chapter 2, a variety of organizational, locational, and managerial factors drive firms to seek international markets, to attempt foreign direct investment in these markets, and to form IJVs to manage this investment. Several models were described in previous chapters which have been devised and used to justify the commitment to IJVs. We will discuss how these models relate to the ventures in the American steel industry to learn more about the industry and about the models. The second primary concern here is to understand how the IJVs studied in Chapters 5 and 6 were managed. Again, the descriptive studies provided here will be compared to conceptual models developed to describe IJVs in a search for similarities and differences. As in the case of formation, this method allows us to understand what is unique about steel IJVs, and what is consistent between these ventures and the broader population of IJVs. Again, we can learn about steel in relation to the general case as well as about the real value of expectations derived deductively from grounded theory to an analysis of a specific situation.

## INTERNATIONAL JOINT VENTURE FORMATION—THEORY AND PRACTICE IN THE AMERICAN STEEL INDUSTRY

The first aspect of IJVs which will be addressed is just why multinational companies would form such organizations in the first place. One of the more interesting developments in these cases relates to a theory which is not addressed in Chapter 2 except as a rejected alternative. This is the Neoclassical Trade Theory with its formal interpretation of the meaning of foreign direct investment. Given all the assumptions of this theory, international exchange should all take place through markets, and all production factors should be immobile. By partially relaxing the assumption of factor immobility, it is suggested that foreign direct investment is the movement of capital from locations which provide low returns to those which provide higher returns because capital is less available in those locations. Foreign direct investment theory concluded that this model was hopelessly unrealistic when Stephen Hymer showed that most direct investment took place among the industrialized nations, all of which were presumed to have plentiful capital.[1] Hymer assumed, therefore, that location-based return on capital investment was not the primary reason for foreign direct investment. Hymer went on to lay the foundation for the input/output (IO) or strategic behavior model of foreign direct investment and, subsequently, for IJVs as well.

The cases discussed in Chapters 5 and 6, however, should give some pause to those who have joined in decrying the relevance of the Neoclassical Trade Theory to the microeconomics of direct investment. Specifically, in every case described here, the U.S. parent sought a foreign partner or an acquisition to provide capital, and the foreign partners all immediately provided large infusions of cash.

This condition reflected a basic imperfection in financial markets. Neither the U.S. equity markets nor the American banking industry was willing to supply the large quantities of capital needed by American steel firms to update their processes in order to remain globally competitive. Over-capacity in the United States and worldwide markets made profits scarce and gave little assurance that large investments would result in large profits. However, foreign capital markets, and particularly the Japanese corporate debt market, were willing to support this massive investment. Japanese banks were not as concerned with levels of profit as they were with the ability to amortize debt. The "patient" capital from Japan was willing to take a long-term perspective on profitability for the various reasons stated in each case; however, that patience proved to have its limits, especially as the Japanese domestic economy experienced a recession. U.S. equity markets, on the other hand, were almost totally driven by current profit levels. In addition, the return to investment in the United States was typically higher than for most investment in Japan; further, the dollar was generally strong during the period of investment. Japanese banks had equity positions in Japanese steel companies. They also held shares in the Japanese auto assembling firms which would provide immediate customers for Japanese quality steel produced in the United States which was provided on a "just-in-time" basis. The Japanese undoubtedly saw that the low quality and productivity of American producers provided an easy opportunity to take market share from competitors, particularly if acquisition led to replacing capacity and not adding to it.

The upshot is that all of the steel IJVs had as a primary raison d'être, the provision of capital to cash-starved U.S. production facilities. Even in two capital rich countries, alternative perspectives on the profit potential of certain industries created a situation where capital from one market could find high levels of demand in the other. While the macroeconomic conditions of the United States and Japan were similar, the microeconomics of the two national steel industries were quite different.

Thus, Asian investment in productive capacity in the United States makes sense from a simple capital mobility perspective. However, the organizational form of foreign direct investment is not directly explained by this slight relaxation of the neoclassical assumptions. Rather than further deconstructing this model, we can look to newer models of industry and organizational economics to explain the details of the transactions.

### Foreign Direct Investment Theory

From the perspective of the eclectic model of foreign direct investment, Asian steelmakers' investments in the U.S. industry would seem to make sense. Ownership capabilities and resources included the access to capital in relation to a neoclassical economics approach to direct investment. They also included the product and process technologies in which the Asian producers were expert as well as the management techniques—total quality, training, team production, and the like—which really turned these plants into competitive facilities. All these concerns were essential. Without foreign control and technology, there was no reason to suspect that the American plants would provide any profits, despite capital infusions. Without the capital, the existence of superior processes made little impact on firms which could not afford them.

Locational factors have already been examined. The actuality and threat of protectionist legislation and regulation made increasing exports undesirable and risky. The Japanese and Korean manufacturers needed to bring their process skills and management methods to the American investments in order to get a better return on their bank capital and to increase sales in the United States. In order to do this, they had to be present in the United States. In addition, as their multinational customers in the auto and appliance industries located in the United States, the need to be close to their customers forced the steelmakers to enter the United States with direct investment. Industry over-capacity made acquisition the efficient method for entry as further increases in capacity could only drive down prices and generate additional political problems.

Finally, internalization was encouraged by the nature of the resources and capabilities brought into the United States. Foreign ownership logically follows foreign investment. The Japanese banks would provide capital in the U.S. market, but not as direct loans to proven problem firms. The capital was available through the Japanese steel companies who could provide the guidance and control to ensure that the American investments would be made efficient and productive. Therefore, the direct investments were made by the Japanese steel companies using bank money provided to them. Ownership came through equity investment. The Japanese bought American assets and became owners of American plants. This ensures both access to profits and the right to control the purchased facilities. Ownership also permitted the Japanese and Korean partners to closely monitor and control the application of their process methods, quality systems, training, and other complex capabilities which were not amenable to licensing the way simple technology might have been. Still, the actual decision to joint venture is not addressed in this model.

### Industry Economics and Acquisitions

The "strategic behavior" model of the joint venture provides our first look at the decision to joint venture.[2] As explained in Chapter 2, this model takes an industrial organization economics perspective on strategic decisions; in other words, it focuses on industry conditions. First among these is the political sensitivity of the steel industry. As the basis for much industrial and military strength, it is of considerable concern to governments. Therefore, the same imperatives which made increased imports risky suggested that complete takeovers of a significant number of steel plants by Japanese firms would arouse government scrutiny and, perhaps, restrictions. Joint ventures, by retaining a local connection, are often safer in a touchy political situation.

Another aspect of strategic behavior is that large multinationals can use joint ventures to subordinate local concerns and control local industry. In this industry, however, foreign steelmakers were neither larger nor more powerful than Americans, although they did have better, more committed management and experience with superior technology. Caves attributes foreign direct investment to superior technological and management capabilities rather than simple market power, and may be close to the mark in this case. Interestingly, it appears that both U.S. and foreign partners expected the Japanese firms, in particular, to provide superior new technology. While this did occur in some cases, in others events it became evident that the technological differences were not great. What the Japanese firms *did* bring to their American investments was superior production management techniques and a commitment to long-term continuous improvement. These capabilities seem to be the basis for the financial successes of the joint ventures studied. However, these skills could not be said to provide dominance to the foreign partners.

Contrary to some industry-focused models, the over-capacity and resulting lack of profits in the U.S. steel industry did not act to discourage foreign investment. However, market crowding and over-capacity in the industry were primary disincentives for greenfield plants. Acquisition, whether whole or in part, changed ownership and management, but did not add to the competition in the steel industry. Of course, some of the excess capacity that drove down profits was created by U.S. imports from these same Japanese and Korean companies. The American ventures replaced trade in finished steel, but were stabilized by imports of raw steel or iron, particularly in the cases of the finishing mills.

A final concern of the industry model of competition is the availability of customers. Customer demand pull from Japanese automakers which were traditional customers of the Japanese steelmakers appears to have been a major factor in attracting steel investment. The auto assemblers were not happy with U.S. quality and were generally uncomfortable with American firms who held critical positions in their logistical chains. By inviting their Japanese suppliers to come to America, these firms provided assurances of initial markets. The same demands for automobile-quality steel from U.S. carmakers provided additional customer pull. In the cases of the West Coast finishing plants, the demand for bar and sheet steel for burgeoning construction and the stability of agricultural demand for can stock made obvious the market potential for lower cost production of the same products.

From an industrial organization perspective, foreign investment in U.S. steel made sense. Political pressure to reduce imports was high. The competition was weak, demand for high-quality and lower-cost production was in place, foreign raw materials provided cost advantage, and the technological and managerial superiority of the Asian producers was apparent. However, while such a market power approach can explain investment, and the attraction of acquisition over new capacity, it does not really justify the use of IJVs. For that, we must look to other models which observe the firms and their individual transactions.

### Transaction Cost Economics and International Joint Ventures

Transaction cost models, such as the internalization model described in Chapter 2, focus on the relative governance costs of different ways of organizing foreign subsidiaries. Market governance, exports in this case, are preferred when markets are free to operate and when specialized investment is minimal. Once markets are disrupted, as by government action, or specialized production is required, as for the automotive industry, multinational firms begin to seek internal hierarchical governance—they build or buy their own facilities in the foreign market. The effect of trade restrictions on direct investment is common to all models, including trade theory. If a firm cannot sell freely from abroad, a natural response is to locate in the target market. The role of specialized investment is unique to transaction cost economics. This economic theory presupposes that partners may be opportunistic and that they will violate the terms of any contract if this option seems lucrative. When one partner is forced to make large investments, such as in added production facilities for specialized steels, then opportunism by the other partner can generate high costs. Thus, to rely only on market transactions, whether with final customers such as the auto companies or with U.S. finishing mills for semi-finished steel, created the risk of being left with an over-capacity if the contracts were violated or not renewed. In such a case, vertical integration by whole or partial ownership can guarantee market access for intermediate products. This appears to have been a significant motivator in the cases of the West Coast finishing mill ventures where POSCO and the Japanese–Brazilian partnership needed outlets for their own semi-finished product.

Another major motivator for internalizing international operations, according to transaction cost economics, is that tacit, organizationally bound, complex knowledge such as process technology and management techniques cannot be adequately transmitted by market relationships such as licensing. For the Japanese steel companies to supply their technology and operations management systems to American plants, they needed direct managerial involvement on a large scale, a situation best solved by bringing these plants inside the Japanese company. This permits the Japanese partner to transmit knowledge through close, ongoing, direct relationships, permits them to put pressure on reluctant partners to adopt new methods, reduces the risk of creating a strong competitor, and ensures a share of the long-term payoffs from organizational change. This condition seems to have been a major motivation for the acquisitions of the integrated operations and of the I/N Tek and I/N Kote joint ventures. Without managerial control, things would not have changed. Without ownership, the necessary control was not available and the rewards for changes in the U.S. plants were not assured.

The question remains: Why did the Japanese not just buy the American plants? Transaction cost models of joint ventures suggest that *both* sides must have difficult-to-market organizational resources and neither side can be able to make an outright acquisition. The Asian companies were not in a position to acquire all of USX, NII, or other American steel companies. However, it is clear that individual plants were available for purchase. (In the case of National, the U.S. partner even preferred this option.) It is also clear that the American companies offered little in the way of useful complex knowledge. Only Inland Steel offered to provide raw materials at a level sufficient to want a guaranteed supply to protect the continuity of highly automated operations. Further, most of the American parents refused to continue infusions of capital. Yet, only in the case of CSI was there not an American parent company involved in the venture. It would seem that the U.S. parents must have offered *some* capabilities to the joint ventures that would be lost in an outright purchase of the production facility. To clarify this point, we turn next to resource-based economics in order to get a slightly different perspective on the way resources and capabilities come together to generate profits.

### Resource-Based Economics and the Steel Joint Ventures

The resource-based view of the firm is like a bundle of resources with some being unique to the firm and some, such as capital, being necessary but openly available in the market. The key to profitability is the possession of unique resources, such as patents, or capabilities, such as management systems, which can be combined to present the marketplace with a unique product. From this perspective, if the most economically rational manner to transmit the firm's unique competencies to a foreign market is through extending the organization, then direct investment will occur. If the unique capabilities of the firm can be enhanced through combination with the capabilities of another firm, then an acquisition or joint venture makes economic sense. In particular, if the resources of the potential partner are tied to the identity of the other firm—its reputation, brand names, national identity, customer or employee loyalty—and would be disrupted or lost to a complete takeover, then a joint venture is the logical way to recombine resources.[3]

As described earlier, the Japanese and Korean partners brought their capital and their skills in process technology and operations management to the U.S. industry. What they found in

American firms was existing plant and equipment (some of which was usable), an experienced workforce, experienced managers with skills in managing American labor forces (although using very different methods), and considerable willingness to accept the obviously more competitive Japanese ways. Both sides brought customer bases including the Japanese auto transplants and existing American markets. Tied to this was the resource that, perhaps more than anything else, drove the desire to joint venture rather than buy plants outright; namely, national identity. The Japanese had preferred the Japanese auto companies because their customers knew that Japanese steel companies would give them the quality and logistical controls that they needed. The American identity of the U.S. partners protected access to possibly chauvinistic American customers and presented a local identity to governments possibly concerned with foreign penetration of a strategic industry.

Events showed that American managers and workers were much more effective than originally thought. Rapid acceptance of TQC methods, discovery that the Americans were aware of the latest technology, and the largely American operating management and workforces confirm this. They were constrained by capital shortages and corporate disinterest in remaining in the industry. However, the cases indicate that most of these discoveries were surprises to the Asian partners. Keeping the American firms in the venture may have also reduced the cost of entry considerably, as the foreign entrants were not forced to buy the existing plants before upgrading them. However, at least in the case of National Steel, the purchase price was minimal. A study by Chang indicates that Japanese firms in the electronics industry prefer sequential market entry in order to reduce risks.[4] A similar dynamic could be in evidence here—exports to joint ventures to (perhaps) whole ownership. It would appear that the primary motivation to retain American partners (CSI, with its Japanese–Brazilian ownership is obviously outside this discussion since Kaiser had completely closed the facility) was to "wrap themselves in the flag," the American flag in this case.

Why would they want to do this? The interest of Congress and the Department of Commerce in protecting the U.S. industry was apparent, but perhaps not very deep. As long as jobs were saved, ownership seems to have made little difference, historically, to the American government. American customers were interested in input value, which meant good quality at the best price. Their embrace of imports indicates that buying an American identity was not a motivating force. If the empirical evidence does not support the need for an American identity, what might? The strong role of the Japanese and Korean governments in building and protecting their home-based steel industries (a foreign takeover would never have been permitted) and their role in pressuring customer industries to buy from home suppliers may have influenced the outlooks of these firms.

Had the roles been reversed, American firms entering Japan or Korea would have been driven by many influences to take on local partners in order to make their entry feasible. It is quite possible that the Asian firms were unable to fully understand the very different attitudes of the U.S. government and the American customers. It is worthy to note that no European steel companies attempted to joint venture in the United States. Did they have no skills of note to offer, or did they see no reason to invest in U.S. facilities so long as they had the right prices?

### Setting Up International Joint Ventures— A Summary

The various economic models appear to suggest that foreign direct investment in the U.S. steel industry made good sense. Threats of import restriction, a desperate need for capital for modernization, and the existing over-capacity of the industry all suggest that foreign acquisitions were economically justified. The case for using joint ventures is less clear from a strictly economic perspective. The macro models offer little assistance.

The organizational economic views of transaction cost and resource-based models accurately predicted that the Asian partners would offer important capabilities that were closely tied to their organizational competencies and required a degree of integration possible only through ownership. They imply that the American sides must have offered similar capabilities to justify using joint ventures while offering little in the way of direct productive value to be derived from retaining the American parent companies as partners. From this, we infer that less apparent economic benefits were expected from retaining an American identity for the acquired plants. The issues of cultural and market unfamiliarity were also identified to have been critical to these decisions.

There may also have been a bandwagon or institutional effect arising from the similar circumstances and the decision by the first entrant to use a joint venture. While not economically irrational, the decisions to use IJVs, as opposed to other means of direct investment, appears to be based more on the social and psychological effects of uncertainty in a novel circumstance.

These effects become even more apparent when one examines the process of actually managing these ventures in the next section of this chapter.

## MANAGING INTERNATIONAL JOINT VENTURES IN THE AMERICAN STEEL INDUSTRY

As a general statement, we find that the six joint ventures described in detail in this book have been able to operate successfully. Only one, the National Steel/NKK joint venture, has "failed." The word fail is used in the sense that one of the partners dropped out of the venture and the operating and financial performance of the IJV was poor. One other, the POSCO/U.S. Steel joint venture, has had noteworthy disagreements between managers from the two partners resulting in a tense relationship; however, financial success kept these issues in the background. Overall, the IJVs have been successful in improving operating efficiencies through effective introduction of new technology and the capital to put it in place. New management systems have been installed in virtually all the plants, and they seem to be successful. The IJVs seem to be gaining market share while steadily reducing their unit costs. Labor relations generally are better than in the U.S. industry as a whole. Conflicts due to organizational and national culture differences appear to be minimal. The problems that have developed seem to come mostly from economic issues relating to the cyclical nature of the industry. Organizational learning has been high within the ventures, but it is less clear that the parent organizations have learned much from the venture association with some notable exceptions. We will examine some of these specific issues more closely, then take a look at the two problem plants.

### Transfer of Technology and Management Systems

One of the primary objectives in establishing these IJVs, in the first place, was to transfer the superior process technologies of the Asian partners to American plants which were operating with obsolete technology. In the integrated mills, National Steel and NKK as well as Kawasaki and Armco had previous experience with each other in R&D partnerships. Kawasaki and Kobe were considered to be among the technological leaders in the industry. These circumstances led to high expectations for technology transfer. However, it appears that the Japanese partners were more important in providing the capital needed to pay for the new processes rather than for the technology itself. While Kawasaki did provide skills in continuous processing at the Lorain plant, the Americans expressed disappointment upon discovering the small technology gap that actually existed between the two firms. Indeed, Kobe's opportunity to learn about pipe-making appears to have been the major technology transfer opportunity in this venture.

Among the finishing mills, neither CSI nor UPI appear to have gained much specific technology from their foreign parents, other than perhaps some improvements in galvanizing and other surface treatments. The Inland Steel/Nippon ventures, however, were specifically created to install superior technology and to supply a specific product to a specific market. In these cases, the two West Coast mills seem to have succeeded in their role of absorbing excess output from the foreign parent and converting this raw steel into marketable products, although it was basically the same product that they had made before. The I/N joint ventures seem to have had great success in installing and managing the new technologies. One reason may be that both parents recognized the importance of these finishing operations. Inland Steel sacrificed some of its own output, recognizing that the consistent quality of new processes would be needed to compete in the automotive market. For Nippon, which supplied the core technology, the new plant seems to have moved beyond their installed technology in Japan, thus giving them a chance to improve their own skills.

While steelmaking technology has had some importance in the success of these mills, management systems seem to be the major contributions of both sides as well as the source of most relevant learning and the primary reasons for their success. In virtually all these cases, operating systems such as total quality control, "just-in-time" logistics, continuous improvement, a focus on preventive maintenance, and extensive training and cross-training of operators were put in place by the Asian-based management and enforced by the foreign owners. Across the board, these systems have brought unit costs down dramatically. However, these systems were not unknown in the United States. They were not resisted in any case, and, in some cases, the American managers took the lead in applying these systems. It appears that ownership which would support and finance such systems was perhaps more critical to the joint ventures than was simple knowledge of their techniques.

The Japanese were seen, in most cases, as obsessed with quality and continuous improvement, which was, apparently, a necessary condition for their institutionalization. On the other hand, the comment from an Inland Steel manager about the difficulty of applying Japanese methods without Japanese direct guidance raises serious questions about the ability of the American parents to learn much which can be taken back to their facilities. The U.S. managers, on the other hand, seem to have provided concepts of business and financial strategies which were new to the Asian partners, and which they seem to be trying very hard to take home. The rotation of foreign managers and engineers will provide them the chance to learn from the joint ventures and carry this knowledge home. The U.S. managers seem to have directed most of their loyalty to the joint ventures themselves. This is not surprising when two of the U.S. partners, USX and NII, had little or no interest in their future in the steel industry. While introduction of Japanese process management systems has been highly successful in the joint venture mills, the opportunity for parent organizational learning seems to be largely ignored by the U.S. parents.

In summary, the role of the IJV in facilitating the transfer of embedded knowledge is supported. This knowledge has had more to do with process management than with specific technical issues. It was already known, at least in the abstract, to the American managers. Nevertheless, it appears that the actual application of Japanese process management techniques was greatly enhanced by direct exposure and top management support.

### Managing in a Cross-Cultural Joint Venture Context

Models of cross-cultural contact in IJVs suggest that cultural differences in organizations will have effects on formation decisions as well as the management of these ventures. Most models, and most experimental results, suggest that greater national cultural differences will lead firms to use contractual or joint venture organizations in preference to wholly owned subsidiaries. Studies of organizational cultural differences indicate that such conflicts increase the difficulty of managing as well as the probability of the failure of joint ventures and partial mergers. In contractual relationships, cultural differences have minimal interaction and in organic expansion, the culture is continuous. In total acquisitions, differences may lead to problems, but these problems can be solved by substitution of the acquiring culture for the acquired. This process may well require driving out critical human resources. Joint ventures and partial acquisitions require working with cultural differences.

The cases examined in this book are all joint ventures and say little about the selection rates of one type of organization over another. However, most models would fairly predict significant cultural problems due to both national and organizational culture differences.

It is interesting to note that we have observed relatively few such difficulties. Most of the ventures reported little internal strife caused by misunderstanding, although language barriers and individual assignment rotation have caused some practical difficulties in maintaining efficiency. In some cases, particularly in the case of NII and NKK, we have observed considerable disagreement between the parents because of very different levels of commitment to the steel industry and, therefore, to the venture. Even in these cases, the different outlooks of the parents did not seem to carry over to the joint venture, although most of the managers were from one or the other of the parents. One hypothesis for the lack of internal strife is that the American managers and workers understood that the alternative to the success of the ventures was a plant closure. They also knew that the American parent company often preferred this option. Also, the focus of the Japanese input was on operational systems which were proven. This was known to the American side conceptually, and had little value connotation. While the Americans saw these systems as obsessive, they fit in a larger context of a growing quality movement and had quick, demonstrable benefits. In addition, many of the American managers involved with the ventures seemed to be eager to try new ways and were happy to be encouraged to do so. Indeed, the small number of Japanese in most cases meant that the Americans executed the installation of the new management systems with training, oversight, and encouragement from the Japanese parent and general neglect from the American parent.

This same situation of Japanese ownership and technical involvement with limited everyday involvement perhaps accounted for the lack of national culture clash. None of the Japanese–U.S. ventures reported significant nationalistic confrontation. CSI had no problems despite dual foreign owners and mostly American management and labor. One reason for general cooperation may be evidenced by the reactions of CSI. The greatest difficulties in understanding were between the Japanese and American steelmakers on the one hand, and the Brazilian iron miners on the other. Apparently, very strong industry institutional norms were at work here and may have carried over into most of the ventures such that all the "steel men" tended to have similar approaches to the business, despite national cultural differences and the idiosyncracies of individual companies. Finally, acceptance of Japan as an industrial culture equal to the United States may have made acceptance of Japanese ideas easier for the Americans.

That this might be an issue is reinforced by the cultural clashes in UPI, where the American managers and workers seemed to have had considerable difficulty dealing with the demands of the Korean parent and its representatives. This was true even though the systems installed by the Koreans were virtually identical to those insisted upon by the Japanese steel companies in their ventures. While Koreans are sometimes considered to be a bit more assertive than Japanese, their cultures are much more similar to each other than either of them are to U.S. culture. The Americans are said to disdain the Koreans, and the seconded Koreans are said to distrust and feel disrespect from the Americans. However, the economics of the company and the success of the infusion of Korean capital and operational control systems have been so favorable that the joint venture has continued to do quite well. It may be that economic success will maintain the relationship until mutual respect is established. On the other hand, the large Korean investment is balanced by the fact that they now have alternative outlets for their raw steel. If USX had more interest in strengthening its position in steel, this mill would be a good prospect for a U.S. buy out now that the new process systems are established. However, USX has not only lessened its commitment to steel, it has totally abandoned any interest in the Western market.

In conclusion, it seems that a variety of economic, industry, and social factors may mitigate the impact of cultural distance in the management of these IJVs.

### Joint Venture Failure in the Steel Industry

The UPI joint venture seems to have had significant problems because of cultural conflicts, but has stayed in business as a joint venture with both parents seeming to be satisfied with the cash flows that the plant has generated. USX has also remained in the industry, even if in a reduced condition. One joint venture mill has had major difficulties, and has failed as a joint venture, despite some movement in a positive direction. This is the NKK/NII joint ownership of National Steel. A close look at this failure is instructive. The most apparent reason for failure was that National Intergroup had no real interest in maintaining National Steel in the first place. NKK would have been wise, in 1984, to anticipate the eventual demise of the partnership as NII was quite clear that it had no real interest in the business and, indeed, had already tried to sell it outright. Undoubtedly, NKK could have made an outright acquisition. If, for the reasons speculated earlier, NKK was not willing to go it alone initially, a planned sequential buy out would have been a wise undertaking. In any event, NKK did acquire a majority interest before NII simply sold its remaining shares in National Steel on the open market, leaving NKK with 75 percent ownership with the remaining 25 percent in the hands of various private investors. In reality, the dissolution of the National Steel joint venture can be seen as a logical process. It was probably in the best interest of both partners. Planning could have avoided recriminations and produced the same results.

Why might this have been wise? NKK was able to install new production equipment, train the workers, institute Japanese-style TQC systems, get wage concessions for employment guarantees, and establish firm customers. However, the remaining influences from the "old" National Steel seem to have been pernicious. Previous management was retained in many cases, but seemed to offer little help in rejuvenating the firm, probably because they believed that their corporate parent, NII, had no interest in their ongoing efforts. Residual labor problems seemed to have remained dormant throughout the period of change, but became apparent when health-care costs began to skyrocket. Eventually, a new American management team was installed; however, this positive event was followed closely by the NII bailout and, at the writing of this book, has not had time to have effect.

At the same time that old problems remained, the firm did make many positive changes. It did, however, have its building financial success interrupted by the 1990–1992 recession in the United States. Just as the extensive, expensive capital improvements and training programs were beginning to pay off, demand collapsed due to macroeconomic conditions. This seems to have been the cue for NII to begin its efforts to drop out of the joint venture. Only improving cash flows would have kept NII around as a passive financial investor. When these dropped, the time had come for them to complete their abandonment of the industry.

For NKK, this period led to considerable frustration, but may have given them a free hand in finishing the modernization of

National Steel. Losing a reluctant, disinterested partner who had no desire to make further investments, must be seen as a positive event. A new management team with no loyalty to the old system must also be seen as positive. Removal of the old owners suggests that carry-over labor problems can be resolved, which is another positive development.

In contrast, productivity improvement had stagnated along with profits at the Lorain mill and USX was only marginally more committed than NII was to National Steel. Nevertheless, dedicated American management and a stronger hand for the Japanese majority owner seem to have provided a positive atmosphere which worked.

## CONCLUSION

These six joint ventures provide many interesting insights on the steel industry and the theory of IJVs. The industry seems to have more life than was expected in the early 1980s. New equipment, new process management, and committed owners appear to make profitable operations possible even in a crowded market. The quality and efficiency introduced by the Japanese and Korean partners appear to have turned these plants around. Even National Steel experienced tremendous improvements which can likely be continued without the anchor of NII dragging behind them. Also of interest are the very different attitudes displayed by the financial communities in the United States and Japan. Lack of capital in a supposedly dying industry contributed greatly to the decline of the American industry. New capital along with a longer term perspective on the economic development of the industrialized world and the role of steel in it have permitted dramatic turnarounds. Properly financed and managed, American steel mills, even integrated operations, could improve their productivity and the quality of their outputs to the point of becoming competitive again.

Joint venture theories receive both support and challenges from these examples. The importance of organizationally embedded capabilities, such as process management systems to the viability of IJVs, is quite evident. Efficiency models, such as transaction cost economics, certainly can step in after the fact and argue that joint venturing is the most efficient way to govern such relationships. This was not apparent at the beginning, however. Concerns for market power or industry dominance were not apparent either. Rather, traditional considerations of capital infusions through equity investment, avoidance of real and potential government interference, improved technology, and better management were the focal issues in these six cases. In addition, the history of joint venturing in the steel industry, although not for final production, certainly provided experience with alliance management.

Cross-cultural issues may have driven the choice of a joint venture as an initial step into a new environment, but also seem to have created minimal difficulties for actually managing these mills, contrary to the speculation of many analysts. Industry culture seems to have been a strong influence in several cases. The "steel culture," openly described in the case of CSI, appears to have had an effect in many cases. As long as the objective was to produce more and better steel more efficiently, the American and foreign owners and managers seem to have been able to embrace new ways, no matter what the source. The only venture to openly disintegrate was the National Steel joint venture, in which the American owner had expressly abandoned its interest in the industry long before it left the venture. Perhaps the one main rule for successful joint venturing is to be sure that both partners actually *want* to be in the business for the long term. No partner seems to be preferable to a partner with no commitment.

Joint venturing was only one tool in the successful rebuilding of the U.S. Steel industry at a reduced but stronger level of capacity, but an effective one. As of mid-1995, no other country threatens its survival. Its productivity is as high and its cost structure as favorable as any. The critical competition is from within, in the form of technologically advancing mini-mills, but they also face rising competition among themselves. More than anything else, the steel IJV experience seems to have proved that marriages among those of common technological languages and industrial commitments can surmount formidable barriers to the benefit of nations as well as firms.

## NOTES

1. Steven H. Hymer, "The international operations of national firms: A study of direct foreign investment." Ph.D. dissertation presented at Massachusetts Institute of Technology (1960). Published by MIT Press (Cambridge, Mass.) in 1976.

2. Bruce Kogut, "Joint ventures: Theoretical and empirical perspectives," *Strategic Management Journal* 9: 319–332.

3. Bruce Kogut and Udo Zander, "Knowledge of the firm and the evolutionary theory of the multinational corporation," *Journal of International Business Studies* 24(4): 625–645.

4. Sea Jin Chang, "International expansion strategy of Japanese firms: Capability building through sequential entry," *Academy of Management Journal* 38(2): 383–407.

# Index


AK Steel Holding Corporation, 77, 96–97, 169
Alliances, strategic, 13–15
  list of, 14
Antitrust policy, 3, 65–66
Armco Steel, 64
  diversification of, 90–91
  joint venture with Kawasaki steel, 67, 90–98, 189
  modernization of, 91–92
Armco Steel/Kawasaki Steel joint venture, 67, 90–98, 189
  management of, 95
  problems with, 94
  results of, 96–98
Automobile industry, and steel industry, 67, 70–71, 85, 91, 96, 103–104, 118, 150–152, 171

Bessemer converter, 30–31
Bethlehem Steel Corporation, 64–65
Brazil, 27, 29, 36–37

California Steel, 67, 92, 169, 185, 190
  and cultural issues, 120–121, 192
  financial status of, 119–120
  investment in, 117–118
  joint venture, 115–123
California Steel *(continued)*
  labor relations, 121–122, 171
  management of, 119–121, 123
  markets of, 118–119
Chicago, 140
China, 37–38
  joint ventures in, 2, 20–22
  as a raw material source, 175
Clinton administration, 138
Close-in delivery, 127
Collective bargaining, 158, 163, 173
Columbia–Geneva Steel Division, of U.S. Steel, 124–125
Compensation, of steel industry employees, 44–50, 58, 147–148, 161, 166
Cooperative Partnership Agreement, 82–84
Corporate takeovers, 57
Cost of living allowance (COLA), 46–47
Cultural issues, 18, 174, 196
  effects on management, 191–193
  and joint ventures, 104–105, 120–121, 130–132, 134, 137, 142, 148–150, 192

Deming circle, 107
Dunning, John, 9–10

Eastern Europe, and international joint ventures, 19–20
Eclectic model of foreign direct investment, 9–12, 182–183
Economic development process
stages of, 25–28
and U.S. steel industry, 28–29, 175–177
Economic models
and foreign direct investment, 9–12, 182–188
and international joint ventures, 15–17, 182–188
Employment figures, of U.S. steel industry, 49
Equity joint ventures, 2, 13–14
formation of, 19–21
risks of, 15
Europe, joint ventures in, 2
European Community, 19
European Union, 6
Experimental Negotiating Agreement (ENA), 46

Finishing mills, 170–171. *See also* California Steel; Inland Steel/Nippon Steel joint venture; UPI
Foreign direct investment (FDI)
and joint ventures, 181–189
models of, 9–13, 182–183
Foreign imports, and U.S steel industry, 40–41, 44–45, 50–55, 100, 160
Franchising, 13
Free trade, 52

General Agreement on Tariffs and Trade (GATT), 138
Germany, steel industry in, 30–31
Globalization, 1, 5
Goldman Sachs, 96–97
Graham, Thomas C., 89, 95, 97
Great Britain, 30–31

Host countries, of international joint ventures, 21–23

I/N Kote. *See* Inland Steel/Nippon Steel joint venture
I/N Tek. *See* Inland Steel/Nippon Steel joint venture
Indiana Harbor steel plant, and Inland Steel/Nippon Steel joint venture, 139–153
Industrialization, 25–29
Inland Steel
history of, 139–140
joint venture with Nippon steel, 67, 93, 141–154
Inland Steel/Nippon Steel joint venture, 67, 93, 141–154
cultural issues of, 142, 148–150
labor relations of, 147–148
management of, 144–145
markets of, 150–151
and plant operation, 142–143
reasons for, 141–142
results of, 153–154
technology and quality improvements of, 151–153
and training of employees, 145–147
Integrated mills, 169–170. *See also* Armco Steel/Kawasaki Steel joint venture; Kobe Steel/U.S. Steel joint venture; National Steel/NKK Steel joint venture
Internalization factors, 11–12, 16
Internalization theory, 15–16, 182–186
International alliance, importance of, 2
International joint ventures, 1–4, 6
advantages for host countries, 21–23
definition of, 12–13
effects on steel industry, 176–177, 196
formation of, reasons for the, 180–189
forms of organization of, 12–13
and global economy, 5–7
Japanese–American, 64–67, 70–72; *see also* Steel industry joint ventures

International joint ventures *(continued)*
managerial reasons for, 19–21
steel industry, 71–74; *see also* Steel industry joint ventures
theoretical advantages of, 15–18
International Organization of Standards, 128

Japan, 27, 29, 33–35. *See also* Steel industry, Japanese
Japanese–American joint ventures, 64–67, 70–72. *See also* Steel industry joint ventures
Japanese management styles, 104–105, 107, 121, 123, 190–191
Joint venture theories, 182–188, 195
Just-in-time delivery, 95, 142, 150–151

Kaiser Steel. *See* California Steel
history of, 115–116
joint venture with Kawasaki Steel, 115–123
Kawasaki Steel. *See* California Steel
history of, 116–117
joint venture with Armco Steel, 67, 90–98, 189
joint venture with Kaiser Steel, 115–123
Kobe Steel, 189
history of, 98–99
joint venture with U.S. Steel, 99–112, 170
Kobe Steel/U.S. Steel joint venture, 67, 93, 98, 170
cultural issues, 104–105
investment in, 100–101
labor relations, 102, 105–106, 108–112
profits, 103–104
reasons for, 99–100
technological and training improvements, 101–103, 105–106
training of employees, 106–108
Korean War, 33, 35
Kwangyang, South Korea, and POSCO Steel, 124, 126–127, 138–139

Labor costs
global, 48
reduction of, 158–159, 161
Labor relations, 10, 172–173
and AK Steel Holding Company, 77
and California Steel, 121–122
and collective bargaining, 158, 163, 173
and Inland Steel/Nippon Steel joint venture, 147–148
and National Steel/NKK Steel joint venture, 81–82, 84–85, 87–90
and UPI, 135–136
and U.S. Steel, 102, 105–106, 108–112
Labor unions, 43–50, 97, 158, 163, 172–173
and Inland Steel/Nippon Steel joint venture, 147–148
and Kaiser Steel, 166, 122
and National Steel, 79, 81–82, 84–85, 87–90
and UPI, 135–136
and U.S. Steel, 108–112
Less developed countries, 6, 21
as steel producers, 37–38, 99, 174–175
Location factors, 10–11
Lorain, Ohio, 98
LTV Steel, 65, 93

Maintenance procedures, steel plant, 83, 121, 151–152
Management styles, 190–191
American and Japanese, 149
American and Korean, 133
Markets, steel, 118–119, 127–129, 150–151, 184
Mini-mills, 42, 57, 59, 66, 99, 162
Modernization, 168, 174

Modernization *(continued)*
of Inland Steel, 141
in Japan, 33
of Kobe Steel/U.S. Steel joint venture, 100–101
of National Steel, 79, 81–84
of UPI, 126
of U.S. steel industry, 67, 70

National Intergroup, 65. *See also* National Steel
National Steel, 49, 168
history of, 78–79
joint venture with NKK, 66–68, 77–90, 93, 168
and USWA, 79, 81–82, 84, 89
National Steel/NKK Steel joint venture, 66–68, 93, 168
advantages and disadvantages of, 80–81
altered operating standards, 84–85
cultural issues, 192
failure of, 88–90, 189, 193–195
financial status of, 85–87
investment, 81–84
labor relations, 81–82, 84–85, 87–90
production, 86
profits, 86
reasons for, 79–80
Neoclassical Trade Theory, 180–181
Nippon Kokan. *See* NKK Steel
Nippon Steel, 34
history of, 140–141
joint venture with Inland Steel, 67, 93, 141–154
Nippon Steel/Inland Steel joint venture. *See* Inland Steel/Nippon Steel joint venture
Nisshin Steel/Wheeling-Pittsburgh joint venture, 67, 93
NKK Steel, 65
history of, 78
joint venture with National Steel, 66–68, 77–90, 93, 168
NKK Steel/National Steel joint venture. *See* National Steel/NKK Steel joint venture

North American Free Trade Agreement (NAFTA), 138

OPEC (Organization of Petroleum Exporting Countries), 41, 46
Organizational models, 17–18
Ownership factors, 10
Oxygen furnaces, basic, 34

Participative management, 105, 110
Pepsico, 19
Performance-based compensation, 108
Pittsburg, California, 124, 137–138
Pittsburgh, Pennsylvania, 140
Poland, 21
POSCO Steel (Pohang Steel Company), 67, 71, 169
history of, 123–124
joint venture with U.S. Steel, 123–139, 170, 175, 191; *see also* UPI
POSCO Steel/U.S. Steel joint venture. *See* UPI
Production, steel, 28, 31–32, 39–41, 45, 59, 160, 164–165
Productivity, labor, 48
Protectionism, U.S., 1, 11, 138, 173, 182
policy, 50, 52, 54, 55

Quality improvements, and joint ventures, 101–102, 106, 128, 151–152, 171

Raw materials
sources of, 175
supply of, 63–64, 174
Reagan administration, 65
Resource-based model, 16–17, 186–188

Salomon Brothers, 96
Smoot-Hawley Tariff, 52
Steel industry. *See also* Steel industry joint ventures
Brazilian, 36–37

Steel industry *(continued)*
Chinese, 73
and globalization, 5
history of, 9
South Korean, 27–29, 35–36
and supply of raw materials, 2–3
western European, 30–31
Steel industry, global, 38
compensation costs, 166
exports, 34
joint ventures, 71–74
production for 1984–1993, 164–165
profits, 53
Steel industry, Japanese
development of, 33–35
and international joint ventures, 73–74; *see also* Steel industry joint ventures
and Kawasaki Steel, 115–116
and Kobe Steel, 95–98
motives for joint ventures with U.S. firms, 66–67, 70–71
and Nippon Steel, 140–141
and NKK Steel, 78
and U.S. market, 54
Steel industry, U.S.
capacity of, 38–41
conditions of in 1984–1994, 157–158
development of, 31–33, 39–42
employee compensation, 44–50, 58, 147–148, 161, 166
employment figures, 49
and foreign imports, 40–41, 44–45, 50–55, 100, 160
future of, 163, 166
and government policy, 50–55; *see also* Protectionism, U.S.
investment in, 10–12, 32, 39–42, 55–60, 67–69
labor costs, 31–32, 34, 43–44, 46, 158–159, 161
labor relations, 172–173; *see also* Labor relations
and labor unions, 43–50; *see also* Labor unions
pricing policy of, 42–44
production, 31–32, 39–41, 45, 59, 160
Steel industry, U.S. *(continued)*
profits, 56–58
and raw materials, 63–64, 174–175
and technology, 157–159, 171, 189–190; *see also* Technology
Steel industry joint ventures, 2–4
Armco Steel/Kawasaki Steel, 92–98
California Steel, 115–123
cultural issues, 104–105, 120–121, 131–132, 148–150, 191–193
effects on steel industry, 196
and environmental issues, 174–175
failure of, 193
Inland Steel/Nippon Steel, 139–154
Kobe Steel/U.S. Steel, 98–112
motives for, 66–67, 70–71, 174–175, 180–189
National Steel/NKK, 77–90
POSCO/U.S. Steel, 123–139
results of, 166
technological advances, 157–159, 189–190; *see also* Technology
and training of employees, 135, 145–147
Strategic behavior model, 15, 183–184
Subsidiaries, wholly owned, 2, 5
Sumitomo Metals Industries, 67, 72
Survival strategies, of U.S. steel industry, 157–160

Technology, 5, 21, 19, 152–153
advances in, 30, 32, 157–159, 189–190
Theoretical models
and foreign direct investment, 9–13, 182–189
and international joint ventures, 15–18, 180–189
Total quality management (TQM), 101, 122, 151
Training, and joint ventures, 135, 145–147
Transaction cost models, 16–17, 184–186

Transnational marriages. *See* International joint ventures
Trigger-price mechanism (TPM), 54

United Auto Workers (UAW), 43
United States, as a foreign steel market, 38, 50–55, 57–58
United Steel Workers of America (USWA), 16, 52, 163, 173
 and AK Steel, 97
 and Inland Steel/Nippon Steel, 147
 and Kaiser Steel, 116, 122
 and Kobe Steel/U.S. Steel, 110
 and National Steel, 79, 81–82, 84, 89
 and UPI, 133, 135
 and U.S. Steel, 159, 173
UPI, 71, 170, 175, 191
 cultural issues, 130–132, 134, 137, 193
 future of, 138–139
 history of, 124–125
 human resources, 133–135
 investment in, 127–129
UPI *(continued)*
 labor relations, 135–136, 172
 management of, 129–131, 133
 markets of, 127–129
 reasons for, 126–127
U.S. Steel, 42–43, 58, 64–65, 71, 193
 joint venture with Kobe Steel, 99–112
 joint venture with POSCO Steel, 71, 123–139, 170, 175
 leadership of, 89
U.S. Steel/Kobe Steel joint venture. See Kobe Steel/U.S. Steel joint venture
U.S. Steel/POSCO Steel joint venture. *See* UPI
USX. *See* U.S. Steel

Voluntary restraint agreements (VRAs), 52, 54–55, 100, 125–126, 173

Weirton Steel Works, 49, 65
Western Europe, 27, 30–31
Wheeling-Pittsburgh Steel, 49
World War II, 3, 31–33

## ABOUT THE AUTHORS

**GARTH L. MANGUM** is Max McGraw Professor of Economics and Management at the David Eccles School of Business, University of Utah. A former steelworker and steel industry arbitrator, with doctorates in both economics and law, he is the author of more than fifty books and monographs and numerous scholarly articles. He has been an economic adviser in more than thirty countries and has served in legislative and executive positions in the U.S. government during the course of two administrations and in an advisory capacity in five others.

**SAE-YOUNG KIM** is Associate Professor and Chairman of the Department of International Trade at the Dankook University, Seoul, Korea. He holds a doctorate in economics and, among other positions, serves on the Board of Directors of the Korean Development Bank and as a consultant to Pohang Steel Company, Korea.

**STEPHEN B. TALLMAN** is Associate Professor of Management at the David Eccles School of Business, University of Utah. He received his doctorate in management from UCLA, and he is the author of numerous articles, conference papers, research projects, and presentations. He is also an editorial board member and a reviewer for various academic journals in international business.

www.ingramcontent.com/pod-product-compliance
Ingram Content Group UK Ltd.
Pitfield, Milton Keynes, MK11 3LW, UK
UKHW041637031026
15587UKWH00001B/349

9 781567 200409